Applied Mineralogy

Technische Mineralogie

Edited by
Herausgegeben von

V. D. Fréchette, Alfred, N. Y.
H. Kirsch, Essen
L. B. Sand, Worcester, Mass.
F. Trojer, Leoben

9

Springer-Verlag
Wien New York 1976

W. E. Brownell

Structural Clay Products

Springer-Verlag

Wien New York 1976

WAYNE ERNEST BROWNELL, B. S., M. S., Ph. D.
Professor of Ceramic Science
New York State College of Ceramics at Alfred University
Alfred, N. Y., U. S. A.

With 120 partly colored Figures

Library of Congress Cataloging in Publication Data. Brownell, Wayne E. Structural clay products. (Applied mineralogy; v. 9.) Includes bibliographies and index. 1. Ceramics. 2. Ceramic materials. I. Title. TP825.B76. 666. 76-40216

ISBN-13:978-3-7091-8451-6   e-ISBN-13:978-3-7091-8449-3

DOI: 10.1007/978-3-7091-8449-3

Preface

Structural clay products have had a place in the history of civilization like bread and cloth. Probably because the industry has been so commonplace in the lives of people, little has been written about it; even the history of its development is sketchy. There is no other book quite like this in publication at present, and it is prompted now because much general scientific knowledge can be, and is, applied to the manufacture of structural clay products. This book is an attempt to bring together in one place the basic sciences that can be useful in all of the processes and experiences of the clayworker.

This volume was written primarily as a text to be used in courses for third and fourth year college students; however, there will be a broader interest in it by industrial foremen, engineers, architects, and scientists employed in the manufacture, research and use of structural clay products. It will also be a source of general information for those interested in entering the field. The treatment of the basic principles of clay products manufacturing and use is so general that even those interested in refractories, whitewares, and pottery may find many parts useful to them.

Structural Clay Products organizes and applies scientific and engineering principles to each production step in the order of its occurrence. It starts with the selection of raw materials and ends with in-service problems and a survey of the present state of the industry in the United States of America. The mathematics introduced has been kept to a minimum and only employed where it can actually be used to solve practical problems or to assist in understanding the scientific principles involved in the various processes. As one goes through this book, it will become obvious that the basic principles of mineralogy are extremely important in the production of high-quality clay products.

I am indebted to many people and organizations for direct contributions to this volume and/or advice, criticism, and encouragement in its preparation. First, I would like to thank my wife, Vivian, for reading and questioning the manuscript; then I must acknowledge the assistance of my technical editor, Dr. Van Derck Fréchette, who made this volume easier to understand and more readable. During the industrial survey at the start of this project, all persons extended to me the greatest hospitality, assistance, and encouragement. The structural clay products industry in the United States is operated by some of the finest people in the world. The following organizations were particularly helpful:

American Olean Tile Company
The Belden Brick Company

The Bonnot Company
Can-Tex Industries, Division of Harsco Corp.
The Fate-Root-Heath Company
General Shale Products Corporation
Glen-Gery Corporation
J. C. Steele and Sons
K-F Brick Company
Merry Companies
National Clay Pipe Institute
National Sewer Pipe Limited
North American Manufacturing Company
Robinson Industries
Swindell-Dressler Company

Alfred, September 1976

WAYNE E. BROWNELL

Contents

1. History and Classification

1.1. Early History

Now the whole earth had one language and few words.
And as men migrated in the east, they found a plain in the land of Shinar and settled there.
And they said to one another, "Come, let us make bricks, and burn them thoroughly." And they had brick for stone, and bitumen for mortar.
Then they said, "Come, let us build ourselves a city, and a tower with its top in the heavens, and let us make a name for ourselves, lest we be scattered abroad upon the face of the whole earth."

Genesis 11: 1—4

The ancients made bricks by hand from materials of the earth which seemed to be naturally suitable. This was some years after the great flood, probably about 4500 years ago, and the people were, no doubt, located in Mesopotamia along the Euphrates River or its tributaries. This area is now part of Iraq. It is clear that the art of brickmaking had already advanced into the process of firing to make them hard, durable, and esthetically attractive. Simple molds were probably used by these people to make the bricks true in shape and size.

Archeological excavations have shown that bricks were simultaneously made in other parts of the world. Walls of fired bricks, similar in shape to today's products, have been uncovered at Monhenjo-daro in the Indus River valley which we now call Pakistan that are about 4500 years old. Sun-dried bricks dating back some 5000 years have also been found in Peru. A Pre-Harappan fired brick from the ancient city of Kalibangan in the Indus River valley is shown in Fig. 1. This conventional-looking product is estimated to be 5000 years old, and it may well be the oldest fired brick on record. For various civilizations in scattered parts of the world to develop in parallel ways the art of brick making, there must be something unique in the properties of clay materials that were easily observed to be useful by these very early people [1].

The making of bricks must extend further back into the history of man, since the state of the art was well advanced by the time of the account in Genesis. Brickmaking probably was the second earliest industry of mankind. Agriculture was the first as man became a cultivator rather than a hunter and gatherer of nature's bounty. The beginning of agriculture gradually changed the behavior of the earliest people

Fig. 1.   Fired brick from the ancient city of Kalibangan, Indus River valley. Courtesy of General Shale Products Museum

from a nomadic life to the start of urbanization. When fertile lands were found suitable for the cultivation of food production, they tended to gather together in these
areas and reside there to till their fields. With some degree of permanency in their
dwelling site, they could then build permanent homes, storage bins for their harvest,
and some protection from invaders who sought to steal their valuable lands. Fortuitously, such locations often contained good flood-plain clay materials for molding
into bricks. It is known that at some locations the making of them preceded pottery,
probably because of the simple shapes of building bricks. The oldest sun-dried bricks
have been found beneath the foundations of the biblical city of Jericho in the Jordan
River valley just a little north of the Dead Sea. Fig. 2 is a photograph of one of these
bricks taken from a structure of a Pre-pottery Neolithic settlement which has been

Fig. 2.   Sun-dried brick from a Pre-pottery Neolithic settlement previous to the
bibical city of Jericho. Courtesy of General Shale Products Museum

carbon-14 dated at 9000 to 10,000 years old. Notice that no molds were used at that time, and thumb-print indentations are clearly visible, put there to effect a key-like bond with the soft, mud mortar used at that time [1].

Around 6000 years ago, rectangular molds were used to form straw-filled bricks at Halaf on the Khabur, a tributary of the Euphrates River in the region which is now Syria. These molds were open top and bottom to allow uniform sun drying. Bricks had not been fired up to this time, and chopped straw was added to plastic clay to prevent cracking during the drying process [2]. There is evidence that at a later date straw was still incorporated to prevent cracking even when the bricks were to be fired; however, the fired brick of Fig. 1 did not contain straw [1].

Decorations and adornments were applied to bricks about 5000 years ago when firing of bricks was introduced. Since then, decorative effects have continually become more complex [2]. Fig. 3 shows a 4000-year-old fired brick bearing an inscription. The message tells us that this brick came from the Temple of Enlil, Goddess of Air, built during the reign of King Ur-Nammu in the Sumerian city of Nippur [1]. This is in the area we now call Iraq. Decorations continued to become

Fig. 3.   Brick from the ancient city of Nippur which is about 4000 years old. Courtesy of General Shale Products Museum

more sophisticated until they reached elaborate proportions in southern Mesopotamia during the period of Assyrian supremacy, about 3000 years ago. "In late Assyrian times glazed brick was often used for decoration and reached its culmination in Babylon after the fall of Nineveh (612–539 B.C.). Figures were modelled in relief on clay slabs; then the slabs were cut into brick sizes by wire before firing and glazing. The famous Ishtar Gate and the walls of the Procession Street leading to the center of Babylon were decorated in this way" [2].

The making of pottery and the use of the potter's wheel came after experience had been gained in the use of clay for brick making. The potter's wheel brought another structural clay product into being. Clay pipes with collar-type joints were made on potter's wheels 3400 years ago to construct drains in the palace of Knossus on Crete [3].

An improvement in brick-mold design had been made by 945 B.C. as is evident from the visible appearance of a fired Egyptian brick shown in Fig. 4. This brick from Bubastis in the Nile delta was made during the reign of Pharaoh Sheshouk I. The lip on the bottom of the brick, as shown, was in reality the top of the open mold

Fig. 4.   Egyptian fired brick from the time of Pharaoh Sheshouk I. Courtesy of General Shale Products Museum

which had sides and a bottom very much like the molds used today on the automatic machines of soft-mud brick production. The lip is left as the excess plastic clay is scraped from the top of the mold.

Although the art and technique of brick and tile making of the Romans was derived from Greek and Etruscan clayworkers, Roman architecture was primarily practical and new construction innovations were developed. The Greeks and Etruscans were adept at making architectural terra cotta and roofing tiles as far back as 1000 B.C. They came to Rome in 496 B.C. to ply their trade. However, about 150 B.C., the Romans discovered that they could make a hydraulic pozzolan cement by mixing volcanic ash with lime. Cement quickly became a major building material in Rome, and brick and tile making ceased in the latter days of the republic. Inasmuch as cement did not make beautiful buildings, facings of brick, tile, and stone were applied. Old bricks and tiles were abundant from previous constructions, and for many years these were used to veneer the new concrete structures. By the time of the Augustan period in the first century, A.D., fired bricks and tiles were made again in Rome for facing official buildings. For this purpose they developed triangular-shaped bricks. A point of the triangle was molded horizontally into the concrete wall; so that the finished appearance was that of a brick wall made from very thin bricks. In addition two-foot square tiles (bi-pedales) were employed as bonding courses at intervals of two or three feet. During this period, private homes were constructed of sun-dried bricks with a stucco of lime, sand, and marble dust applied to the exposed wall surfaces [4].

During the first century, the Romans transported their art of brickmaking to Great Britain. The Roman legions brought with them their own brick and tile makers. An interesting tile product was made by them which was a part of the heating system for buildings in the cooler climate. Fig. 5 shows a Roman hypocaust tile from Bath,

Fig. 5. Roman hypocaust wall tile from Bath, England. Courtesy of General Shale Products Museum

England made in the first century. A network of these decorated, hollow tiles was set in the walls of a room beneath which was a furnace. The hot air rose through the tiles providing heat and comfort for the room [1].

It wasn't until the 13th century that the English began making bricks by their own techniques, but this started a very stable industry which has persisted to modern times. The importance of roofing tiles to the people of Great Britain was emphasized by the Parliament of Edward IV. In 1477 they set down rules for the manufacture of these tiles. It seems that some good often comes from tragedy. The great fire of London in 1666 gave a great impetus to the brick industry. Within a few years the city was transformed from a town of wood to a city of brick, and a safer city it was [5]. By this time the British had exported the art of brickmaking to America.

That the early American colonists lost no time in making use of the brick clays is shown by the numerous records which mention the erection of brick kilns at many locations in the 17th century. Bricks were made first in Virginia in 1611. Iron-shod molds were used by the Massachusetts colonists for the production of bricks in 1629. That the industry flourished is shown by the fact that in 1643 the watch house at Plymouth was built of brick, and land in Malden was sold in 1651 to a brickmaker named Johnson whose operations were evidently extensive, because 18 years later laws were passed to keep the clay pits from encroaching on the highways. (This historical note has some overtones of present-day problems in it!) The quality of bricks produced in Massachusetts was regulated by the courts in 1667. Governor Stuyvesant introduced brickmaking into New York City about 1646. An indication that brickmaking was close to the life of the people was found in W. L. Stone's History of New York City published in 1872. He noted that in 1661 breweries, brick kilns, and other manufactories carried on successful businesses. Note the possible order of priorities [6].

The early settlers in Maine started making bricks in 1635, and they were the first to export their products—to the West Indies from the Port of Piscataqua in 1789. Bricks were made in North Carolina in 1663, in Rhode Island in 1681 and in Pennsylvania in 1683. Brick production in America followed closely the first settling of an area, and this tendency followed the migration of pioneers into the West [6].

The first flat roofing tile was made in Montgomery County, Pennsylvania about 1735 and in Bethlehem, Pa., in 1740, but the industry did not assume a permanent character until machinery had been invented for mass production.

The nature of the brick industry in the United States up to 1800 was well described in a little story by Chute [7]. Wood-fired brick kilns along the Quinnipiac River in Connecticut were operated as a part-time occupation by farmers who made bricks at odd moments during the season, which closed in the early fall, with the burning time set as a social festivity during October, when the irresponsible services of the owner's friends and intoxicated volunteer helpers often threatened the results of the entire season's work. The capacity of the largest kilns was from 20,000 to 75,000 bricks; occasionally an owner burned up to 100,000. Even with this relatively small quantity, many manufacturers were from 3 to 5 years in disposing of their ware, so slow were the sales.

1.2. Industrial Revolution

A great change took place in the production of structural clay products, as in all manufacturing, immediately after 1850. To fully comprehend the nature of this sudden change, we must go back to the latter part of the 18th century in England. It was here that a "revolution" was born which was to affect the economic and social structure of the people in many parts of the world, including the United States. For 10,000 years the major sources of economic wealth had been agriculture and commerce, and the making of things did not make an equivalent impression on society. In the 15th century Alberti advised his readers on the best ways of getting rich. They were to engage in wholesale trade, seek a hidden treasure, ingratiate oneself with a rich man to become his heir, loan money, or to rent pastures, horses and the like. There was no mention of making things for sale. It is true that in the 2nd century a Roman brickworks employed 46 foremen, but such manufactures were exceptional, and they were not the same economic enterprises as were to spring up after 1850 [8].

The actual beginning of the industrial revolution, as far as the structural clay products industry is concerned, was with the invention and development of the steam engine. James Watt worked long and hard on the development of an unsatisfactory engine invented by Newcomen, and finally came up with an extraordinarily powerful and efficient one in 1796.

Inventions alone would not have started the industrial revolution without some way to produce and sell the machines conceived. Eventually, Watt was fortunate in teaming up with a wealthy and highly successful manufacturer by the name of Matthew Boulton, and by 1786 they created much excitment in and around London when they harnessed two steam engines to 50 pairs of millstones and created the largest flour mill in the world. The demand for steam engines for various applications mush-roomed immediately thereafter, and many "new" men (men who were the first of a new social class) became very wealthy [8].

In the period from 1790 to 1850 steam engines were improved and adapted to the mechanization of many manufacturing processes. The increased capital required and the accelerated output of goods required elaborate factories to be built which in turn required factory workers with new kinds of jobs. Thus, the economic and social revolution was well underway. It was during this period that the first steam powered, soft-mud brick machines were invented in the United States. One of the early machines was introduced by Henry Martin and used around Perth Amboy, N. J.; however, it took about 30 years for such machines to become universally accepted, and in the meantime the bricks were hand molded as had been done for centuries [6]. In 1859 John Craven in England invented a steam powered stiff-mud extrusion machine for the forming of bricks [9]. Stiff-mud machines were also in use in the United States by 1860. A machine with an automatic cutter was put into operation by the Chambers Brothers Company at Pea Shore near Camden, N. J., about 1862 [6].

A necessary complementary component to the brick machines was the Hoffmann continuous kiln developed in Europe in 1859 and introduced shortly thereafter into the United States. This complex structure took care of the firing of the increased output from the forming machines, and great fuel savings were made over that re-quired for the previously used periodic kilns [9].

The use of steam power also provided stronger mineral dressing equipment, so that grinding machinery for argillaceous shales became available. This opened up vast new stores of raw materials for use by the structural clay products industry. Here-tofore, only conventional clay deposits had been used throughout history. One of the first to use shales for the making of bricks was L. G. Eisenhardt at Horseheads, N. Y., in 1880 [6].

History is cloudy on the introduction of machine power to the dry pressing of structural clay products. This is because mechanical, hand-operated presses were in use as early as 1829 in the United States, and they were gradually improved over the years. It is known that some presses were hydraulically operated by 1856 and powered by steam in 1870. The manufacture of building bricks by the drypress process was short lived in any case, because of the introduction of extrusion equipment. The extrusion of end-cut brick was so efficient that pressed bricks could no longer com-pete in the market place. Presses have continued to be used in the structural clay products field for forming floor and wall tile.

The industrial revolution really took hold of the structural clay products industry near the end of the 19th century when machinery had been developed for all phases of production and when the traditional resistance of a very ancient art would tolerate change. Steam shovels were needed for mining large quantities of clay and shale to feed the high-speed forming machines. Power driven grinding and mixing equipment was necessary for the preparation of the raw materials for plastic-forming operations. Fans were required for drying and firing the formed ware, and power was required to move large masses of products through the kilns.

With machinery available for the integrated operation of structural clay products plants, products other than building bricks started to be produced in the United States on an industrial scale. In 1872 some streets in Charleston, W. Va., were paved with paving bricks. This part of the industry grew rapidly in the late 1800's, only to disappear again in the 1920's almost as fast as it started. Plank reported that the first interlocking roofing tile plant was started in Terre Haute, Ind., in 1871. By 1891, red roofing tiles were made at Alfred Center, N.Y., and in 1892 plants were started in Montezuma, Ind., and Chicago Heights, Ill., by the Ludowici Company. Eventually, the roofing tile business found better markets in California [10]. The manufacture of partition tile for the construction of fire-proof buildings was started in 1875 in New Jersey. Also, by 1875, the small plant using hand molds and the potter's wheel to produce sewer pipe was a thing of the past. Steam presses and extrusion techniques had become the heart of the sewer pipe industry [3]. Glazed or enameled bricks, previously imported from England, were produced at Momence, Ill., and Mount Savage, Md., in 1893, and by 1896 other plants were in operation at Saylorsburg, Pa., and Sayreville, N.J. In 1875, the American Encaustic Tile Company was organized at Zanesville, Ohio, to make floor tiles. After a slow start, success was achieved, and by 1800 a line of glazed wall tiles was added. By 1890, fourteen other factories were established in the eastern states to produce floor and wall tiles, both glazed and unglazed. The 1870's were the years of impact of the industrial revolution on the manufacturing of structural clay products in the United States [6].

After a few premature starts, the production of architectural terra cotta was started in 1849 as the Hall Terra-Cotta Works at Perth Amboy, N.J. In 1879, this business was incorporated under the name of the Perth Amboy Terra-Cotta Company [6]. This early terra cotta was made from red-firing and from stoneware clays. Examples of the red terra cotta manufactured in the 1890's are shown in Fig. 6. The complex decorative features of architectural terra cotta are brought out in the photographs. The excellent durability of these products against the destructive forces of weathering is dramatic when one considers that these pieces of terra cotta shown here have been exposed to the weather since about 1895, and are still exposed today. It is unfortunate that this industry did not stay with these raw materials and methods of manufacture.

The first forty years of the industrial revolution in the United States produced such a rapid growth in this very ancient industry that a need was recognized for standardizing the sizes and shapes of bricks made in the various factories throughout the land. For this reason, the American Society for Testing Materials was established in 1898 to bring agreements among producers and customers on such matters. Later this

Society became the general authority on specifications for quality and durability of all other types of construction materials.

Fig. 6. Unglazed, red terra cotta manufactured at Alfred Center, N.Y., in the late 1890's. Courtesy of D. E. Rase

Although several Hoffmann kilns were erected in the United States, only a few were used by the structural clay products industry. This was probably due to the introduction in 1903 of a continuous, straight-line tunnel kiln by the United States

Brick Company at its Oaks Factory in Phoenixville, Pa. This kiln was 576 feet long and probably used coal for fuel. This turned out to be an important technological advance for the industry in this country, and this type of kiln was destined to become universally accepted. Throughout the 20th century, the tunnel kiln has been refined and perfected for use by the structural clay products industry.

In the 1920's automatic tile presses were being developed, and they were made and used by the Franklin Tile Company of Lansdale, Pa. With the combination of automatic presses and continous kilns, a one-fire, straight-line production of floor and wall tile was put into operation in 1929 by this company. Such a high-speed, efficient operation has brought this business into prominence among the various structural clay products industries [11].

An extremely important technological advance in the forming of extruded clay wares was being introduced in 1932. This was the incorporation of a vacuum chamber through which the plastic clay was forced to flow immediately before it was pushed through the shaping die. This technique increased product strength in all stages of production (wet, dried, and fired) and improved uniformity. Vacuum extrusion healed the laminations inherently put into the products by the auger which was a necessary part for continuous extrusion machines. This principle is universally used in the industry today.

Although the steam engine was the impetus for industrializing and advancing the structural clay products industry from hand-made production to mechanization, this power source did not last long. The mechanized steam powered plant came to this industry about 1860, but by 1900 electric motors were operating the equipment in the plant, and by 1920 many internal combustion engines were being used for mining and for transportation equipment both in and outside the factories.

A great change was induced in the structural clay products industry starting around 1925 when the mechanization in the cement industry placed on the market large quantities of less expensive concrete products. The paving brick industry declined very rapidly in the middle 1920's with the introduction of concrete for streets and highways. The peak in production of common bricks for backup units in wall construction was reached in 1923. Afterwards, concrete blocks, poured concrete, and steel became the basic materials for load-bearing walls and internal partitions, and as in Roman times, bricks were relegated to the role of facing material. Ceramic hollow tiles for fireproof partitions within buildings gave way completely to cement products so that by 1952 the production of these tiles had dropped to a negligible value.

The structural clay products industry has now learned to live with the competition of cement products. The brick industry has essentially recovered from the loss of the common brick market by greatly expanding the production of facing bricks. The major products of the structural clay products industry today are facebrick, sewer pipe, floor tile, and wall tile. There are, however, small quantities of paving bricks, floor bricks, structural tiles, drain tiles, roofing tiles, and conduit still being produced in the United States. Note that no mention was made here of the architectural terra cotta industry. This story will be taken up later.

During the industrial revolution in the structural clay products industry, factories appeared at a rapid pace wherever the population concentrations created a market. Some of the enterprises were well founded and others were shaky because of poor management and technical incompetence. Such a haphazard state of affairs led inevitably to mergers and consolidations of previously separate businesses. This started in 1899 when the National Fireproofing Company of Pittsburgh took over plants in Pennsylvania, Ohio, New Jersey, Massachusetts, and Maryland. In the same year, the Baltimore Brick Company consolidated plants in its area. The New England Brick Company of Boston also bought up plants in New England in 1899. The next year found The Illinois Brick Company of Chicago taking over plants in that vicinity. In 1905 the Ludowici and Celadon companies of Cleveland merged and controlled plants in several states [6]. The driving forces of consolidation in this period were to extend successful management, to regulate prices for better profitability, and to better plan production to fit the market demands.

For similar reasons, certain companies in the industry continued to grow during the years of prosperity up to the 1930's. The Gery brothers, already experienced in the brick industry, founded a new brickmaking business in 1908; they acquired other factories in 1916, 1920, 1925, and 1937. In 1928 the Kingsport Brick Corporation and the Johnson City Shale Brick Corporation in Tennessee merged to form the General Shale Products Corporation. This corporation, within two years, bought plants in Richlands, Va., and Knoxville, Tenn.

Not all of the mergers of this period were successful as pointed out by Kirk in 1932 [12]. A merger of 8 plants in Iowa and about 30 plants in other states failed after two years. This unfortunate experience led Kirk to question the benefits to be gained by consolidations of this magnitude in this industry; however, he suggested the mergers were not well planned to accomplish specific goals. This analysis of the goals to be achieved, even in today's industrial-economic climate seems to be an essential consolidation procedure. It is necessary to stress this point because such mistakes are still being made in the structural clay products industry.

Other driving forces toward the rapid industrialization of the structural clay products industry in the first forty years (1860–1900) of the industrial revolution were the changes in population and the urbanization of society. The population of the United States more than doubled in this period from 31,513,000 in 1860 to 76,094,000 in 1900 [13, 14]. Immigration was adding millions to our population every year. All of these new people required homes and substantial buildings in which to work. During this period urbanization was in full swing, and fireproof cities were proven to be a necessity. This rapidly rising demand took the manufacture of bricks, tiles, and pipes out of the hand-working processes. It was a necessity to use mechanical, high-speed equipment which, in turn, called for well-organized factories and capital to use and obtain the necessary facilities. It was no longer possible for farmers and tradesmen to keep up with the expanding economy.

By taking the stable brick industry as an example of the experiences of all segments of the structural clay products industry, one can see the effects of economic disturbances on the economic health of this industry. Fig. 7 shows the variations in brick production in the United States during the 20th century in relation to population as a measure of potential demand. The solid line traces production over the years, and the dashed line represents population growth in the same period. It might be generally expected that the production of bricks would increase with the demand imposed by increasing population; however, the catastrophic effects of wars and

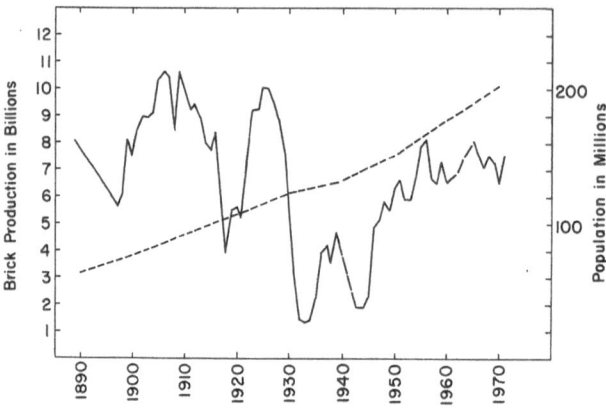

Fig. 7. Relation between brick production in the United States and population [13, 14]

economic depressions on this industry are clearly reflected in the figure. The financial panic of 1907 made itself felt in the 1908 production of bricks. World War I caused a short-term decline in construction which was compensated for by a spurt in brick production in 1922. About 10 years after the war, the economic depression of the 1930's reduced brickmaking to an extremely low figure relative to the needs of the population. Just as construction and brick production were starting to recover from the economic collapse, World War II practically put a halt to brick production for four years. Since the end of World War II, the brick industry has been straining to meet the demands. With the steadily increasing population, the structural clay products industry can expect to play a catch-up game after every period of economic instability.

1.3. Scientific Revolution

We can look back on history and see the start and development of the industrial revolution and its impact on our culture. Now we are in the midst of another socio-economic change which C. P. Snow [15] has called the "scientific revolution", but its limits cannot be clearly seen, since we are on the inside looking out. The effects of this revolution can only be predicted at this time, but it is extremely important for us to realize that it is happening and that it will influence our culture perhaps more than the introduction of machines. The impact of the scientific revolution has already touched the structural clay products industry, and it is a force that must be recognized, more clearly than at present, by this industry.

C. P. Snow places the beginnings of the scientific revolution in the 1920's or when we began to make practical use of subatomic particles. For the structural clay products industry, the discoveries which led up to this new change were made somewhat earlier than that. By 1935 the scientific knowledge adaptable to this industry was available, but the depression and World War II delayed the application of it until after 1946. This was too late for the architectural terra cotta part of the industry. Even now, the whole industry is slow to adapt itself to the scientific revolution. Unlike a new industry created by the change, the traditions of thousands of years seem to be hard to break.

As far as the structural clay products industry is concerned, the seed of its scientific revolution was planted in 1895 by W. K. Röntgen when he discovered X-rays. This seed was dormant but germinating until 1912 when Laue and Friedrich brought forth the first sprout when they showed that X-rays were diffracted by the atoms in crystals. W. H. Bragg [16] and W. L. Bragg [17] tenderly nurtured this sprout with their works on the mathematical solution of crystal structures from X-ray diffraction data in 1913. These plants of the scientific revolution grew randomly and haphazardly from one place to another as many scientists produced X-ray patterns of every crystal they could get their hands on. Diffraction pattern after diffraction pattern was added to the technical literature. Of special interest to us was the application of X-ray diffraction to such fine-grained materials that even microscopes could scarcely detect their crystallinity. These were the substances found in soils and clays previously thought to be amorphous.

In 1930, L. Pauling [18] reaped the harvest of all this scientific cultivation, and the basis for the scientific revolution in the structural clay products industry was set. He brought together all the scattered information on the flat, mica-like minerals and put it in order. From this, the crystal structures of the clay minerals were deduced, and for the first time, the structural clay products industry knew what it was working with. Previous to this development, the clay minerals had defied precise description because of their ultrafine crystal size.

Further advancements in the science of clay minerals, which are the foundations of all clay working, were immediately forthcoming. Hofmann, Endell, and Wilm reported in 1933 on the structure of a particularly interesting clay mineral called montmorillonite [19]. Ross and Hendricks provided additional extensive description of this mineral in 1945 [20]. Brindley told the industry in 1951 how to identify the clay minerals and all of the rest of the mica-like minerals related to them [21]. Then, Grim put together the clay mineral structures with their properties in water, drying behavior, and products formed on heating. This he did in his book first published in 1953 and revised in 1968 [22].

Another scientific tool was adapted for use by the structural clay products industry which allowed examination and identification of crystalline matter in both raw materials and finished product. This was the petrographic microscope. The ceramic community of our country was encouraged to make use of it by T. N. McVay [23] in 1934. This instrument is still used extensively for examining the microstructure of clay products upon which many physical properties depend.

All of ceramics is involved with high-temperature chemical reactions. A scientific breakthrough was made by H. L. LeChatelier in 1884 when he studied the exothermic and endothermic reactions on heating clays to high temperatures [24]. To measure these heat effects, he used a Pt-Pt10%Rh thermocouple. Little use was made of his work by our ceramic industry until Granger applied the principle of a differential thermocouple to measuring the heat effects on heating clay minerals in

1934 [25]. Insley and Ewell [26] used this technique, called differential thermal analysis, and Norton [27] evaluated the method as applied to clay mineralogy just prior to World War II. Differential thermal analysis proved to be a very powerful tool, especially in conjunction with X-ray diffraction, for the purpose of understanding the firing process and improving firing practices. This technique began to be used industrially after World War II.

Electronic instrumentation for laboratory research and plant operation became available to the structural clay products industry after 1946 as a result of the electronic circuits developed during the war. Rapid progress has been made in this area since then, right up to the present time. The developments of cathode ray tubes, ceramic transistors and magnets, and the efforts of the space exploration programs have provided many new instruments and refinements of others for the guidance and control of industrial processes, including those of the clay products industry.

Recently, an acoustical instrument was developed by G. Robinson at Clemson University which allows one to listen to and record the sounds of microcracking and macrocracking occurring inside a product during firing and cooling. This technique was immediately put to use by the industry [28]. This laboratory instrument will be of great help in efforts to fire and cool at optimum rates with an expectation of savings in fuel and improved products.

About the time that the need for technically trained people to support the industrial revolution was recognized, the scientific revolution came along and demanded them. As industrialization gradually became more complex and people expected more from it, the inventor and handyman were no longer adequate. Engineers were required to build, operate, and maintain factories. The practical use of science, which started the scientific revolution, created so much excitement that more and more scientific principles were turned to profit. At the end of World War II, nearly everyone in this country believed that the solutions to all problems were to be found in basic scientific research.

The governments of the highly industrialized states began to provide engineering education in the ceramic fields around the turn of the century, even before the industries themselves fully recognized the need. In the beginning, if a factory never had employed an engineer, it could not relate to the necessity. Ceramic engineering education in the United States was started in 1894 by Edward Orton, Jr. at Ohio State University. The present New York State College of Ceramics was founded in 1900, and its first leader was Charles F. Binns. The third university to set up an engineering degree program in ceramics was Rutgers in New Jersey. This school was established in 1902 and was directed by C. W. Parmelee. The next two state universities to enter the field were Illinois in 1905 and Iowa in 1907. Today there are 20 state universities offering engineering and science programs in ceramics and many other universities have courses in ceramic science to support other majors.

That the structural clay products industry was slow to pick up the services of college-trained engineers and scientists, is apparent from the comments by Richardson [29] in 1903. He observed that the lack of technical knowledge in the industry was holding back its profitability. He analyzed the situation with these words, "We now undoubtedly lead the world in the efficiency of our brick-making machinery, but in accurate technical knowledge of clays, and the best methods of treating them for special purposes, in a scientific system of drying and burning, in fact in all the chemical processes involved in the manufacture of clay products, we have much to learn, and there is only one way to learn it, and that is by the training of our young men in technical schools, giving them a thorough course in ceramic engineering." This appeal was, for the most part, ignored by the structural clay products industry, and in 1934 this provoked Norton [30] to say, "The ceramic industry needs more of the highly trained engineers and scientists than it realizes. . . . As an example of the need for thoroughly trained men, let us consider the tile industry. It is surprising that it has allowed itself to be handicapped by the troublesome crazing of glazed tile. It is beyond question that noncrazing tile can be made. . . . All that was required was a thorough knowledge of the principles involved and a more careful control of the manufacturing process.

"As another example, consider the field of face brick. This industry has lost millions of dollars worth of business because so many leaky walls have been installed. . . . This industry has also lost a great deal of business from efflorescence on brick walls, which in many cases is so unsightly as to deter the architects from chancing the use of this material. While at the present time we do not thoroughly understand the cause and cure of this trouble, . . . it would not seem to be an impossible task to find the cause and cure. Here again the industry needs technical men with enough vision to see these problems in their broad aspects and to solve them before other types of material have made serious inroads on the prospective market."

For thirty some years, while the urgency for the application of scientific principles to the structural clay products industry was being pointed out, a branch of this industry was collapsing for lack of scientific principles. The early architectural terra cotta expanded rapidly into at least seven states by 1910. Soon after the manufacture of these highly ornamental building products was started, there was an increasing demand for color. Slip or engobe coatings were widely used around 1884 for the application of color, and glazes were becoming commonplace by 1900. This change in market demands led the industry away from the red-firing bodies toward buff-firing stoneware clays and low-grade fireclays with 20% to 45% grog added to control shrinkage. This lighter-colored body was more suitable for the application of bright colors [6, 31]. This change of materials and the application of glazes led to trouble.

Even as the architectural terra cotta industry continued to expand, problems in weathering durability were occurring. The glazes were crazing in service, cracks in the body were developing, and spalling of the decorative surfaces was completely ruining some installations. Such weathering problems were being noticed by 1915 and Clare worried about them in 1917. He listed the causes of these difficulties as defective body mixtures, frost action, glaze defects, and irregular firing conditions [32]. In 1922, Hill thought the failures might be due to the admission of water to the terra cotta through poor mortar joints, poor wall construction, and moisture expansion of the product in service caused by the iron content of the body [33]. As now, it was typical of the industry to try to pass the blame on to causes further removed from the manufacturing processes or on to gremlin-like demons that only the devil could control.

Because the industry lacked the scientific personnel and programs to cope with their problems, the National Bureau of Standards began to study the terra cotta problem in 1917 [34] under a fellowship sponsored by the National Terra Cotta Society. The first breakthrough to understanding the problem causing disaster in the industry came in 1927 when Spurrier measured the delayed moisture expansion of certain common ceramic bodies [35] which caused glaze crazing and ultimate destruction of the product by the freezing action of the water entering the body by way of the cracks in the glaze. Two years later Schurecht and Pole found the remedy for moisture expansion without understanding the basic causes and effects [36]. They reduced moisture expansion by the additions of magnesite or blast furnace slag. They also found that it was reduced by firing the bodies to low porosity so that water could not penetrate.

Unfortunately, these efforts were too little and too late. The architectural terra cotta industry had expanded to a peak around 1926 even while their products were failing in many locations. A new tunnel kiln plant called the "world's best plant," was opened in 1926 by the Northwestern Terra Cotta Company of St. Louis [37] and there were at least 27 plants in this industry about this time. The catastrophic failures reached disastrous proportions by 1931, and the industry was collapsing by 1935, unable at this late date to apply the findings of the researchers working on the problem. The American Ceramic Society dropped the Terra Cotta Division in 1945, and only a few plants continued to operate up to 1960. By this time the market for architectural terra cotta was gone. As far as is known, there is only one plant producing custom-made terra cotta at this time, and it is in California. This deplorable experience should have awakened the whole structural clay products industry to the necessity for basic and applied research for a better understanding of their raw materials, processes, and service behavior well in advance of critical need.

The scientific revolution taking place in this country finally began to penetrate the structural clay products industry by 1949. Because the industry was composed of a few large companies and many small ones, efforts bent toward cooperative research and development programs. In this year both the National Clay Pipe Institute and the Structural Clay Products Institute embarked on extensive research and development programs. About this time 4 or 5 large companies instituted their own quality control, and applied-research programs. All of these efforts were probably stimulated by the research programs of several universities which had been feeding useful concepts to the industry since 1929.

The National Clay Pipe Institute established a research laboratory in 1949 at Los Angeles. Work of general interest to the industry was carried on there until 1957 when the laboratory was moved into new quarters at Crystal Lake, Illinois, where it is still very active. The value of this research program to the industry is recognized by the innovations and improvements that have come to the industry since 1950. A better understanding of raw materials and processes have brought unglazed vitrified pipe, plain-end pipe, factory fabricated compression joints, and more profits to the member companies.

The Structural Clay Products Institute, composed of members manufacturing facebricks and structural tiles, established the Structural Clay Products Research Foundation at the Armour Research Foundation in Chicago under the direction of Robert B. Taylor. Work was started there in 1950 to attempt to reduce the cost and improve methods of construction of brick and tile structures. The research program expanded rapidly, and it was necessary to build a separate research facility at Geneva, Illinois, in 1955. A most active and successful research program, broad in scope, was carried on for the next 13 years at the Geneva facilities and at the New York State College of Ceramics at Alfred University. Unfortunately, the member companies still did not understand the need for research in order to keep their businesses productive and prosperous. They became impatient (in less than 10 years) with the rate of research progress; they were unwilling to change their factories in order to use the results of research; they could not agree on what lines of research should be undertaken. Also during this period, a few large companies were becoming larger, and some major corporations were buying into the industry for diversification. These organizations lost interest in the cooperative venture, since they could finance their own research and development programs better tailored to their individual needs. As a result, the research effort was essentially abandoned by the Institute in 1967 when new quarters were built at McLean, Va. A small program in construction design and testing is still being carried on there. (The organization has recently changed its name to the Brick Institute of America.)

Even during this brief tenure of research in the brick and tile industry several significant accomplishments were made. A trend toward different sizes and shapes of products was established. Blast-resistant construction design was made and tested under conditions of nuclear explosions. A vast program to collect data on the thermal values of various types of walls was put together. The factory-fabricated panels of brickwork were initiated here. Automatic packaging of products for shipment was developed. The troublesome, in-plant problem of scumming of products was solved, and the service problem of efforescence became thoroughly understood. It was regrettable that other research programs underway could not have been continued.

Some of the significant accomplishments of the industry in adapting itself to the scientific revolution since 1960 are:

1. Improved drying and firing through better control instrumentation and equipment design.
2. Automatic hacking of products from the forming operation onto tunnel kiln cars.
3. Packaging of products for shipment.
4. Improved control and adaption to raw materials.
5. Completely durable products more universally produced.

At the present time the automatic unloading of kiln cars is being worked out at a few locations, and everyone is conscious of fuel savings and possible conversions from one type of energy source to another.

A truly significant experiment in factory design and operation was tried by the General Shale Products Corporation at Knoxville, Tenn. This came out of a cooperative research and development program undertaken in 1965 by General Shale Products Corporation, Acme Brick Company, Bickerstaff Clay Products, Boren Clay Products, and Harrop Ceramic Service Company. A completely automatic plant which fired facebricks without stacking was built by General Shale in 1967, and it operated as such until 1974 when modifications were made. There were some difficulties in the firing process, but this basic idea is not dead. Technically, it can be accomplished now with the experience gained from the General Shale Products initiative.

As a result of the scientific revolution in the industry and the growth of population in the United States a new driving force for the acquisition and mergers of companies made itself felt around 1950 and is still continuing today. The cost of research, automatic equipment, instrumentation (including computers), new tunnel kilns, dryers, and marketing has resulted in the need for greater capital assets than are available to small companies. From 1952 through 1970, many small privately owned plants were bought by larger, public-owned corporations and mergers were made with other related industries.

There are other forces at play and advantages to be gained by the expansion of companies in the structural clay products industry. Good management can be extended. More flexibility in the product line produced at any particular plant can be gained, and better product distribution can be achieved with a more efficient sales force.

We are now in the scientific revolution, and only speculation will give us an idea of where it will lead us. One can be certain that it will be good for the structural clay products industry and the quicker the industry takes advantage of the evolving scientific knowledge the better. Some speculation as to future trends in this industry will be taken up in the final chapter.

1.4. Classification of Structural Clay Products

Numerous times during the historical development mention was made of one type or another of structural clay products. Before the scientific and technical aspects relating to this industry are pursued in greater detail, it is appropriate to view at a glance all of the products which come under the title "structural clay products." To facilitate this perspective, they have been put into a kind of order in Table 1. Four distinctly different types of products have been subdivided according to variations resulting from manufacturing methods, end use, or accepted terminology.

As indicated in the title of this table, there are color variations in all products listed. The basic colors are red, buff, and white, depending on the starting raw materials; however, there will be found many shades of these colors caused by compositional variations and firing conditions. The reds may grade from pink to almost black and the buff colors from light cream yellow to neutral gray. Body stains are sometimes used to promote brown, blue, green, and black. There seems to be no end to the number of color variations on facing bricks and tiles and on floor and wall tiles which can be deliberately created by coatings, textures, and glazes for esthetic effects. One thing that becomes apparent, however, is the necessity to control the colors within the limits of human visual perception.

Also indicated in the title are numerous dimensional variations deliberately imposed by various sizes and shapes of these products. For example, bricks may vary from 11 5/8 × 3 5/8 × 1 5/8-inch pavers (29.5 × 9.1 × 4.1 cm) through the standard, 8 × 3 5/8 × 2 1/4-inch, brick (20.3 × 9.2 × 5.7 cm) to through-the-wall units that may be 11 5/8 × 7 5/8 × 3 5/8 inches (29.5 × 19.3 × 9.1 cm) or larger. The number of different sizes and shapes in structural facing tile is even greater, and everyone has seen the variations appearing in floor and wall tiles. Clay pipes vary from 4 to 42 inches in diameter (10.2 to 107 cm). Excessive dimensional variations in any of these products are not permitted. Constructional units of all kinds have to fit!

Table 1. *Classification of Structural Clay Products**

I. Bricks
    A. Facing
       1. Extruded–stiff-mud process
          a. Solid-plain, textured, or glazed
          b. Cored-plain, textured, or glazed
          c. Panel-plain, textured, or glazed
       2. Sand-molded–soft-mud process
          a. Sand-textured
          b. Glazed over sanded surface
       3. Drypressed
       4. Lightweight
    B. Common
    C. Paving

II. Tile
    A. Structural (for construction of walls)
       1. Partition
       2. Facing–plain or glazed
          a. Regular
          b. Acoustical
          c. Lightweight
          d. Through-the-wall
       3. Chimney
    B. Floor
       1. Quarry
       2. Large–(various shapes) plain, textured, or glazed
       3. Mosaic
    C. Wall–usually glazed
    D. Flue
    E. Roofing

III. Pipe
    A. Sewer and drain–plain or glazed
       1. Hub-end
       2. Plain-end
    B. Chemical resistant–plain or glazed
    C. Conduit–plain or glazed
    D. Drain tile
       1. Plain
       2. Perforated

IV. Architectural Terra Cotta–plain or glazed

* Color and dimensional variations exist within each product listed.

Bricks are primarily subdivided into end-use categories such as facing, common, and paving. Each of these general types of bricks require special properties suitable to their intended use. Today, common and paving bricks are a small part of the total brick production. Common bricks are not expected to be used for facing where they would be subjected to severe weathering conditions. They are most often available as off-grade ware, a result of production mistakes or accidents; however, some companies are prepared to make them on order for firewalls inside of buildings and the like. Of course, such bricks sell at lower prices than regular facing bricks.

The subclassifications of facing bricks are made on the basis of manufacturing processes. Products made by the extrusion of plastic clay material through a shaping die are called stiff-mud products in the industry. Bricks made by this method can be solid or cored if a bridge is placed in the die. Panel bricks are made in appropriate sizes and shapes to be used in the making of prefabricated panels which are then put in place as a completed portion of a wall. Sand-molded bricks, made by the soft-mud process, are formed in damp wooden molds lined with sand to prevent sticking of the clay. Although in today's factories this is a highly automated process, it is an adaptation of the pre-industrial method of making bricks by hand. Facing bricks are not made in the United States by the dry-press method at this time, but it is included here because it is a process that may come into favor again, since technology has advanced a great deal from the time the method was abandoned. A fourth subclassification of lightweight bricks is placed in this table because it has been a product of interest to the industry for 15 or 20 years. Although the processes proposed to date have not met with favor, such products may appear on the market in the not-too-distant future. Whether they will be made by extrusion, molding, or pressing cannot be known at this time.

The classification of tiles is based on end use rather than on method of manufacture. All except thin floor and wall tiles are formed by extrusion. Plastic and dry pressing are the methods used to produce floor and wall tile. Structural tiles are used for the construction of load bearing and nonloadbearing walls in buildings. Only a few partition tiles are made today because of the more common use of cement products for primary wall construction. These tiles are only used to construct walls and partitions in buildings where some kind of facing or covering is to be applied. There are many varieties of facing tiles all of which are designed for the construction of walls with a decorative effect on the face so that on completion the wall will immediately have a finished surface. Acoustical structural tiles are also decorated, but they have holes drilled through the face into the cavities for the controlled absorption of sound. A through-the-wall tile has both faces finished, width suitable for a complete wall, and close dimensional tolerances. The two sides may be decorated in different ways to accommodate architectural preferences in different areas. Chimney tiles, sometimes called bricks, are employed to construct the circular walls on the outsides of tall industrial stacks. Floor and wall tiles are essentially selfexplanatory, and they are made in several designs, shapes, and colors. Quarry tiles are made in various shapes, and they are so vitreous that they do not require a glaze. Most of the sewer pipes manufactured today are unglazed, and there seems to be a tendency toward the production of plain-end pipe. Drain tiles are pipes used to control underground waters in agricultural fields, along highways, and around buildings.

Three things may be inferred from this brief description of structural clay products. *First*, some form of quality control in the manufacturing operations is essential to maintain color, size, and shape within acceptable limits. The importance of quality control is underscored in Chapt. 9, and by a close examination of Fig. 8. The collage

Fig. 8. Collage representing various structural clay products. (Designed by Rita Elaine Powers)

not only shows the varieties of structural clay products but brings out the esthetics involved and the degree of color and shape uniformity that is required. The human eye is a critical instrument for detecting slight variations in color and even slight variations that may occur in blends of colored units. *Second*, research should be carried out to improve the performance qualities and to keep the costs competitive with other materials. *Third*, engineering should continually be directed toward better factory design and operation.

References

1. Brick . . . Its History and How it is Made, General Shale Products Museum, Johnson City, Tenn.
2. van Oss, J. F.: Materials and Technology, Vol. II., Longman-H. H. DeBussy, Amsterdam, pp. 218—21 (1971).
3. The Story of Clay Pipe Yesterday and Tomorrow, National Clay Pipe Manufacturers, Washington, D. C.: 1958.
4. Encyclopaedia Britannica, 19 (1959).
5. Ratner, M.: The Clay Products Industry in Ohio, National Youth Administration in Ohio, Occupational Study No. 2, (1938).
6. Ries, H., and Leighton, H.: History of the Clay-Working Industry in the United States. New York: J. Wiley and Sons. 1909.
7. Chute, A. H.: Marketing Burned-Clay Products. Ohio: The Ohio State University, Columbus. 1940.
8. Heilbroner, R. L.: The Making of Economic Society. Englewood Cliffs, N.J.: Prentice-Hall. 1962.
9. Anonym: The science and art of brickmaking—Part I. The story of the brick—and historical survey. Claycraft 35, 414—17 (1962).
10. Plank, R. D.: Origin and manufacture of California's clay roofing tile. Am. Ceram. Soc. Bull. 13, 180—83 (1934).
11. Bell, F.: Notes on a 50-Year Revolution. Lansdale, Pa.: American Olean Tile Company. 1973.
12. Kirk, R. E.: Are mergers a benefit to the heavy clay industry?, Part II. Brick and Clay Rec. 81, 169—70 (1932).
13. Statistical History of the United States from Colonial Times to the Present, Fairfield Publishers, Stamford, Conn. Distributed by Horizon Press, New York, N.Y.: 1957.
14. U.S. Bureau of the Census, Statistical Abstract of the United States: 1948—1972, Washington, D. C.
15. Snow, C. P.: The Two Cultures: And a Second Look. New York: Cambridge University Press. 1963.
16. Bragg, W. H.: X-ray reflection by crystals. Proc. Roy. Soc. A. 89, 246—48 (1913).
17. Bragg, W. L.: X-ray diffraction and crystal structure. Proc. Roy. Soc. A. 89, 248—77 (1913).
18. Pauling, L.: The structure of micas and related minerals. Proc. Nat. Acad. Sci. U.S. 16, 123 (1930).
19. Hofmann, U., K. Endell, and Welm: Kristallstruktur und Quellung von Montmorillonite. Z. Krist. 36A, 340 (1933).
20. Ross, C. S., and S. B. Hendricks: Minerals of the Montmorillonite Group, U.S. Dept. of the Interior, Geological Survey, Prof. Paper 205-B (1945).
21. Brindley, G. W., ed.: X-ray Identification and Crystal Structures of Clay Minerals. London: The Mineralogical Soc. 1951.
22. Grim, R. E.: Clay Mineralogy. New York: McGraw-Hill. 1953, 2nd ed. 1968.
23. McVay, T. N.: Identification of crystalline substances by means of the petrographic microscope. Am. Ceram. Soc. Bull. 13, 255—60 (1934).

24. Le Chatelier, H.: De l'action de la chaleur sur les argiles. Bull. Franc. Mineral. **10**, 204–11 (1887).
25. Granger, A.: Thermal analysis of clay. Céramique **37**, 58 (1934).
26. Insley, H., and R. H. Ewell: Thermal behavior of the kaolin minerals. J. Res. Nat. Bur. Stds. **14**, 615–27 (1935).
27. Norton, F. H.: Critical study of the differential thermal method for identification of clay minerals. J. Am. Ceram. Soc. **22**, 54–64 (1939).
28. Edwards, J.: Thermal acoustical analyzer helps solve cooling problems. Brick and Clay Rec. **165**, 27–29 (1974).
29. Richardson, W. D.: The work of the ceramic engineer in the brick-making industry. Trans. Am. Ceram. Soc. **5**, 237–41 (1903).
30. Norton, F. H.: Ceramic education. Am. Ceram. Soc. Bull. **13**, 150 (1934).
31. Wilson, H.: Monograph and bibliography on terra cotta. Am. Ceram. Soc. Bull. **5**, 94–145 (1926).
32. Clare, R. L.: Causes of failure of terra cotta in the wall. Trans. Am. Ceram. Soc. **19**, 593–96 (1917).
33. Hill, C. W.: Terra cotta problems suggested for discussion and investigation. J. Am. Ceram. Soc. **5**, 732 (1922).
34. Powell, W. H.: Address on terra cotta. Am. Ceram. Soc. Bull. **3**, 255 (1924).
35. Spurrier, H.: Some observations on terra cotta physics. J. Am. Ceram. Soc. **10**, 686–92 (1927).
36. Schurecht, H. G., and G. R. Pole: Effect of water in expanding ceramic bodies of different compositions. J. Am. Ceram. Soc. **12**, 596–604 (1929).
37. Anonym: From backyard shop to world's best. Brick and Clay Rec. **68**, 698–702 (1926).

2. Mineralogical Composition of Structural Clay Products

2.1. Structure of Disilicate Minerals

When clay materials are tempered with water to produce the plasticity necessary for proper workability, it is the clay minerals that are responsible for these properties. To understand the interaction between clay minerals and water, it is necessary to know the crystal structures of the various clay minerals. In addition, many of the disilicate minerals other than those of clays are often found in, or added to, the raw materials for structural clay products. These other disilicate minerals are of more concern in the firing operation than in the development of plasticity, but the crystal structures of these minerals, as well as the clays, is important in following the progress of the high-temperature reactions. The structural clay products engineer and scientist

Fig. 9. Tetrahedral coordination of oxygen ions (transparent) about a silicon ion (opaque)

cannot solve the technical problems that arise without a good foundation in the crystal structures of the disilicate minerals.

The clay minerals belong to the family of *disilicate* minerals. All of them are characterized by a silicon:oxygen ratio of 2:5, and this ratio is expressed as an $(Si_2O_5)^{2-}$ unit in the writing of formulas. In general appearance the disilicate minerals usually are flaky like mica. In fact, the mica minerals are a part of this group. The crystals are *micaceous* because of the layered arrangement of the ions in their structure, especially the strongly bonded silicon-oxygen sheet.

In every silicate structure the silicon ion is surrounded by and bonded to four oxygen ions as shown in Fig. 9. This is a tetrahedral arrangement, i.e., if one draws lines from the center of one oxygen ion to another, thereby connecting all four, the solid figure so generated is a 4-sided figure, a tetrahedron. When we put the four oxygens (transparent spheres) together in this way, there is an unoccupied space in the center of the tetrahedron. The size of this space depends on the radius of the oxygen ions. It is just the right size to accommodate the silicon ion; therefore, it seems the geometry of cations and anions determines, at least in part, the *coordination* or groupings in which they occur.

With this *silica tetrahedron* as a building unit, silicate structures are derived by having neighboring tetrahedra share one oxygen ion, i.e., oxygen serves as a common

Fig. 10. Silica tetrahedra sharing a corner

corner of two tetrahedra. This type of sharing is visible in Fig. 10 where two silica tetrahedra are joined. In this case, one corner of the lower tetrahedron is shared with the upper one. Now, if all four corners are shared with other silica tetrahedra, a 3-dimensional network is formed, as is the case in a *quartz* crystal where the formula is SiO_2, and the silicon to oxygen ratio is $1:2$. However, when this sharing is extended only in two dimensions instead of three, a hexagonal pattern is generated as can be seen in Fig. 11. This arrangement can be extended indefinitely in two dimensions

Fig. 11. Hexagonal structure generated by six silica tetrahedra sharing corners

forming a sheet of silicon and oxygen ions. This suggests a flaky, mica-like appearance if extended to a macroscopic scale.

Most common clay minerals are essentially aluminum silicates. The Al^{3+} cation is somewhat larger than the Si^{4+} cation; therefore, it prefers to be surrounded by six oxygen ions instead of four. The packing together of six oxygens forms an 8-sided figure, an *alumina octahedron* in which the central void space is larger and more appropriate for the aluminum ion. This configuration is shown in Fig. 12 where the six transparent spheres represent oxygen anions and the small opaque sphere is proportional to the size of an aluminum ion.

When octahedra combine, they often share two oxygens instead of one as in the case of tetrahedra. Thus one octahedron shares an edge with its like neighbor. This arrangement can be seen in Fig. 13 where the two transparent oxygen ions in the center (upper front and lower back) are common to both alumina octahedra; this is the shared edge. As in the case of silica, this sharing can be extended in three dimensions. This produces a corundum (a-Al_2O_3) crystal; however when the shar-

Fig. 12. Octahedral coordination of oxygen ions about an aluminum ion

Fig. 13. Alumina octahedra sharing an edge

ing takes place only in two dimensions a hexagonal configuration is again formed as Fig. 14 shows. Notice that where octahedra are so assembled, there is an unoccupied

Fig. 14. Hexagonal structure generated by six alumina octahedra sharing edges

octahedral site in the center. This vacancy is necessary to preserve the electroneutrality of the positive and negative ions. If you wish to calculate this balance from Fig. 14, remember that some of the oxygen ions shown will be sharing with other cations not shown as the configuration is extended. There is a similarity of this 2-dimensional structure to that of the silica sheet, again reminiscent of the flaky appearance of the mica minerals.

Magnesium is also associated with the chemistry of the clays and other mica-like minerals. Because the ionic size of the Mg^{2+} cation is similar to the Al^{3+} cation, it also occurs in edge-sharing octahedral configurations. If these octahedral units are shared in three dimensions, a crystal of periclase, MgO, results. Fig. 15 shows the hexagonal appearance that results when magnesia octahedra are shared in only two dimensions. This configuration is like the 2-dimensional alumina sheet with one exception; the central octahedral space is now occupied by a magnesium ion. The additional magnesium ion is required in order to maintain electroneutrality because of the 2^+ charge on the magnesium ion instead of the 3^+ charge of the aluminum ion.

Fig. 15. Hexagonal-appearing structure generated by six magnesia octahedra sharing edges

Hydroxyl anions, $(OH)^-$, play a similar role to the oxygens in the disilicate mineral structures. The hydroxyl ion can substitute for oxygen because the sizes of the two are nearly the same. This substitution takes place only with the octahedrally coordinated aluminum and magnesium ions–never with the silicon ions. As a matter of fact, when the substitution of hydroxyls for oxygens is complete, two platy, layered minerals are formed, namely *gibbsite*, $Al(OH)_3$, and *brucite*, $Mg(OH)_2$. The crystals of these minerals are derived by placing the 2-dimensional sheets on top of each other like the pages in this book. The clays and the other micaceous minerals all have some substitution of hydroxyls for oxygens in the octahedral sheets previously described.

Perhaps by now it has struck you, as it did Pauling [1], that a whole family of disilicate minerals could be derived by stacking up 2-dimensional sheets of silica, alumina, and magnesia in various ways so that elements of the silica tetrahedral sheets would share elements of the octahedral sheets whether they are aluminum or magnesium. Indeed, such is the case. A host of disilicate minerals, all having layered structures and a flaky appearance, are built from silicon-oxygen sheets combined with either a gibbsite-like sheet or a brucite-like sheet. Those combining silica sheets with octahedrally coordinated alumina sheets form a group of layered minerals called *dioctahedral* disilicates. The dioctahedral term in itself describes the alumina octahedral sheet where two out of every three possible positions for aluminum ions are filled, as was pointed out on Fig. 14. When silica sheets are combined with octahedrally coordinated magnesium sheets, the minerals generated are called *trioctahedral* disilicates indicating that all three of the positions just referred to are filled with magnesium ions as was shown in Fig. 15.

Essentially, there are two ways in which the silica sheets can be superimposed on the octahedral sheets. Some minerals are formed by placing one silica sheet on one octahedral sheet to form one layer of a new mineral. These minerals are termed *2-layer*. Others are made by placing a silica sheet on both sides of the octahedral sheet to create a different type of layer called *3-layer*. This is a symmetrical arrangement like a sandwich with a slice of meat between two slices of bread.

2.2. Classification of Disilicate Minerals

This order of things seems to suggest that the disilicate minerals could be put into a tidy classification. This we will do with some reservations. Certain substitutional modifications induce complications when attempting to put each mineral into its own pigeonhole, but a classification is presented in Table 2 for the benefit of clarity, keeping in mind the difficulties in classification. The table first arranges the disilicate minerals in three parts: 2-layer, 3-layer, and regular mixed-layer. These categories are quite sufficient for minerals of importance to the structural clay products industry. Secondly, in each major category the minerals are divided between dioctahedral and trioctahedral except in Part C where the minerals are all trioctahedral.

The simplest and purest of the clay minerals leads off the classification. *Kaolinite* is the combination of one silica sheet with one gibbsite-like sheet to form one layer of kaolinite. The way in which these sheets are combined can be seen in Fig. 16 where the silica sheet is on top and the gibbsite-like sheet is on the bottom. The large, dark spheres represent hydroxyl ions, and the clear spheres still represent oxygen ions. In kaolinite, the silicon ions are found to be surrounded by four oxygen ions, but each aluminum ion is coordinated with four hydroxyls and two oxygens. The layer is unsymmetrical, with oxygen ions on the top and hydroxyl ions on the bottom. As the layers are stacked one upon the other, the hydroxyl groups then form hydrogen bonds between the layers. This interlayer bonding is, of course, relatively weak, but still it is stronger than the Van der Waals bond between unassociated molecules. In order to effect the hydrogen bonds between the layers of kaolinite, it is necessary to adjust one layer upon another so that the hydroxyl ions of one layer are in closest proximity to the oxygens of the other layer [2].

Table 2. *Classification of Disilicate Minerals*

A. 2-layer Minerals—1 silica sheet + 1 octahedral sheet

1. Dioctahedral minerals

kaolinite	$Al_2(Si_2O_5)(OH)_4$
halloysite	$Al_2(Si_2O_5)(OH)_4 \cdot nH_2O$

2. Trioctahedral minerals

chrysotile	$Mg_3(Si_2O_5)(OH)_4$
greenalite	$Fe_3{}^{2+}(Si_2O_5)(OH)_4$

B. 3-layer—2 silica sheets + 1 octahedral sheet

1. Dioctahedral minerals

pyrophyllite	$Al_2(Si_4O_{10})(OH)_2$
beidellite [3]	$X_{0.3}^+ Al_2(Al_{0.3}Si_{3.7}O_{10})(OH)_2$
montmorillonite	$X_{0.3}^+(Mg_{0.3}Al_{1.8})(Al_{0.3}Si_{3.7}O_{10})(OH)_2$
nontronite	$X_{0.3}^+ Fe_2{}^{2+}(Al_{0.3}Si_{3.7}O_{10})(OH)_2$
illite [4]	$K_{0.6}(Mg_{0.2}Fe_{0.3}^{2+}Al_{1.5})(Al_{0.6}Si_{3.4}O_{10})(OH)_2$
muscovite [5]	$KAl_2(AlSi_3O_{10})(OH)_2$

2. Trioctahedral minerals

talc	$Mg_3(Si_4O_{10})(OH)_2$
saponite	$X_{0.4}^+ Mg_3(Al_{0.4}Si_{3.6}O_{10})(OH)_2$
vermiculite [6]	$Mg_{0.3}(Mg, Fe^{3+})_3(Al_{1.2}Si_{2.8}O_{10})(OH)_2 \cdot nH_2O$
biotite	$K(Mg, Fe)_3(AlSi_3O_{10})(OH)_2$
phlogopite	$KMg_3(AlSi_3O_{10})(OH)_2$

3. Regular, mixed-layer, trioctahedral minerals
Solid solutions of the chlorite group
1 biotite-like layer + 1 brucite layer
general formula: $(Al, Fe^{2+}, Mg)_6[(Al, Fe^{3+}, Si)_4O_{10}](OH)_8$

        amesite
        corundophyllite
        prochlorite
        clinochlore
        pennite

The appearance of real crystals of kaolinite is quite suggestive of the models so far shown. Fig. 17 is an electron photomicrograph of kaolinite crystals magnified several tens of thousands of times. The very thin crystals are somewhat transparent as can be seen where one crystal lies above another. The hexagonal form is derived from the hexagonal array of cations which was seen in the models. The crystals of kaolinite are not of the hexagonal system, however, because the stacking arrangement of layers to produce those hydrogen bonds causes the crystallographic axis *c*, in the direction of stacking of layers, to be at an angle to the hexagonal-appearing plates. In fact the kaolinite crystal is triclinic.

There are two other polymorphic forms of kaolinite, *nacrite* and *dickite*, but these are of no concern to the structural clay products industry. Kaolinite is a very important mineral to the industry, since it occurs in or is added to these ceramic compositions. The purest form of kaolinite-bearing clay is a residual kaolin, often called china clay.

Halloysite is a layered mineral, but it occurs in the form of hollow tubes. This mineral is sometimes found mixed with kaolinite as in some of the North Carolina kaolins. Since these tubular crystals cannot grow large, halloysite-containing kaolins are more plastic. In the formula for halloysite, n is a number from 0.5 to 1.5 and represents water molecules on interlayer positions. This mineral is of little concern to the structural clay products industry except where special clays are selected for floor and wall tile bodies.

Fig. 16. Ionic model of kaolinite

Notice the structural formulas as they are written in the classification table. They are useful because they give so much information at a glance. The (Si_2O_5) unit indicates that the mineral is a disilicate with a layered structure and will most likely be micaceous in form. Since there are five other types of silicates, it is good to make this distinction immediately. Further examination of the formula tells whether it is a di- or trioctahedral mineral and whether it is 2- or 3-layer. The rules for writing the structural formulas of silicates are important to fully appreciate their implications. These rules are: 1. All of the cations occurring in 4-fold, or tetrahedral, coordination are enclosed in parentheses together with the oxygen ions. Sometimes, as we shall presently see, cations other than silicon are in tetrahedral coordination, and if they are they are also included in these parentheses and counted as silicon ions when determining the Si-O ratio. The reason for this is that these other cations are playing the role of a silicon ion in the structure. 2. All of the cations existing in 6-fold, or octahedral, coordination are written immediately to the left of the tetrahedral parentheses, like Al_2 in the kaolinite structural formula. If more than one element appears in octahedral coordination, they are all enclosed in the same parentheses. 3. If there are larger cations or hydrated cations which require larger coordination numbers than 6, they are written to the left of those in octahedral positions. In the disilicate minerals, such ions will be found between the layers, where their coordination number is most likely 12, i.e., six anions from each layer. 4. Finally, the hydroxyl ions are written to the extreme right-hand side of the formula.

To write a proper structural formula, one needs to know only the structure of the mineral. If the mineral is a one-to-one sheet disilicate, the writing of the formula is started by putting down the characteristic disilicate group, $(Si_2O_5)^{2-}$; then the octahedral group is placed to the left. If it is a dioctahedral mineral, Al_2^{6+} is written there; however, if it is trioctahedral, Mg_3^{6+} is placed there instead. Now the number

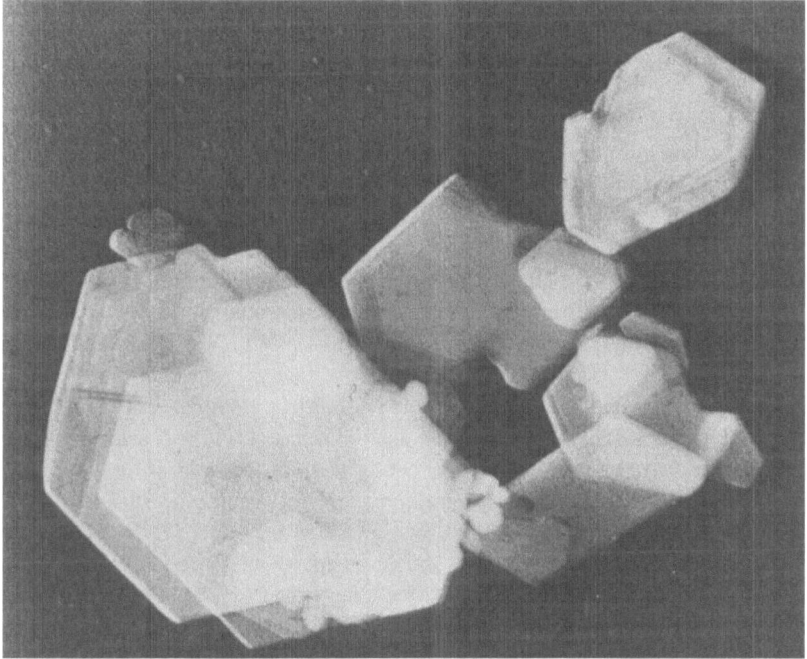

Fig. 17. Electron photomicrograph of kaolinite crystals

of hydroxyl ions is readily determined by the number of negative charges necessary to balance the formula electrically. In both cases it needs to be 4; so the formula for a 1:1 layer dioctahedral mineral is written as for kaolinite in Table 2, and the formula for a 1:1 layer trioctahedral mineral is constructed as for chrysotile.

With these rules one can write structural formulas without even knowing the names of the minerals expressed, and one can read all of these things from a prewritten structural formula. They are very expressive notations in silicate chemistry.

The 2-layer, trioctahedral minerals of Table 2 belong to what is known as the *serpentine group*, which seems to be an appropriate name because a variety of chrysotile is commonly known as asbestos, and we all know how fibrous this mineral is. These minerals are of little concern to those involved with structural clay products, but they are included here for continuity. In addition, there is a substitutional variation shown here which is important to recognize because such substitutions occur later in minerals of much greater importance. The mineral *greenalite* has divalent ferrous ions in place of magnesium ions. This substitution is possible because the size of the Fe^{2+} ion is nearly the same as the Mg^{2+} ion.

The 2-layer dioctahedral minerals are extremely important in structural clay products, and they present another behavior pattern in disilicate mineralogy. Several clay minerals are derived here by cationic substitution from one end member to another. These end-member minerals are pyrophyllite and muscovite.

If one writes the structural formula for a pure 3-layer, dioctahedral mineral, it turns out to be *pyrophyllite*, an important nonclay mineral used in the production of floor and wall tiles. Notice that the two silica sheets, one on each side of the octahedral sheet, is written as (Si_4O_{10}), and since silicon can only be associated with oxygen ions, the number of hydroxyls is cut in half when the formula is balanced electrically.

Next after pyrophyllite in the classification is a group of three similar clay minerals derived when a small and random substitution of Al^{3+} ions for Si^{4+} ions occurs in the tetrahedral sheet. This substitution is allowed because the size of the aluminum ion is actually intermediate for a perfect fit between octahedral and tetrahedral oxygen coordination; therefore, it is found in both coordinations depending on the environment in which the crystal is formed. In very general terms, aluminum ions occupy octahedral sites in silicate structures when a crystal is grown at low temperatures and tetrahedral sites at higher temperatures.

This substitution of Al^{3+} for Si^{4+} in clay minerals creates a positive charge deficiency-which is compensated for by the introduction of interlayer cations. In the *beidellite* formula the interlayer cations are expressed as X^+ because they are variable and are exchangeable in a water environment. They could be di- or trivalent as well, but they are shown here as monovalent cations to simplify the observation of electroneutrality.

When a small amount of Mg^{2+} ions substitute for Al^{3+} ions in the octahedral sheet, a charge deficiency does not necessarily occur, since there is a vacant octahedral site in the dioctahedral layer, and three magnesium ions equal the charge of two aluminum ions. In effect the dioctahedral sheet becomes partly trioctahedral. This substitution occurs in the common clay mineral *montmorillonite*, which is the major mineral present in bentonite.

Nontronite is another clay mineral with properties similar to montmorillonite and beidellite, but it is distinguished by the ferrous ions in the octahedral layer. These three minerals have similar properties, and they are sometimes called part of a *montmorillonite group* of minerals. The interlayer cations are exchangeable in a water medium because their random occurrence on interlayer positions near the sites of charge deficiencies apparently promotes weak bonding. Some cations geometrically fit the interlayer sites better than others; so the interlayer bonding of the exchangeable cations is variable and weak.

When potassium occurs as the interlayer cation, it fits the holes in the hexagonal structure very well, and it becomes nonexchangeable. This modification plus a larger substitution of aluminum ions in the tetrahedral sheet produces the very important, red-firing, clay mineral, *illite*. This is probably the most widely distributed clay mineral in the world, especially in the temperate latitudes and on the ocean floor. Many structural clay products are made from materials where illite is the major clay mineral.

Finally, when the aluminum ion substitution in the silica sheet becomes regular and there is one aluminum for every three silicons, the other end member of the 3-layer, dioctahedral group is reached. This mineral is *muscovite*, a mica mineral with potassium ions bonding the layers together [5]. This arrangement provides for the strongest interlayer bond of all the minerals discussed. Muscovite is a very common mineral in clays and shales; therefore, it is found in small quantities in most structural clay products raw materials.

Like the dioctahedral minerals just described, the 3-layer, trioctahedral group also has end-member minerals in a substitutional series, and there are also several important minerals here for structural clay products. The pure end-member mineral is *talc* which only differs from pyrophyllite in having a magnesium octahedral sheet instead of an alumina octahedral sheet in this position. Talc is only of interest, so far, in the floor and wall tile industry where it may be a prepared-batch ingredient. As with the other 3-layer group, a small random substitution of Al^{3+} for Si^{4+} in the tetrahedral sheets creates a mineral which appears to be the trioctahedral analog of montmorillonite. As a matter of fact, *saponite* is a trioctahedral clay mineral with properties similar to montmorillonite, and there are, at least, two other trioctahedral clay minerals often listed here, but they are all of little concern to the structural clay products industry.

Vermiculite is a highly substituted form of saponite, but its exchangeable inter-layer ions are only hydrated magnesium ions. This arrangement causes an indefinite number of water molecules to be associated with the magnesium on interlayer positions. It is this interlayer water which produces the exfoliation on rapid heating [7]. Vermiculite is not used as an additive to structural clay products at this time, but sometimes it is found in crude clays and shales used by the industry.

At the end of the substitutional series when the substitution of Al^{3+} for Si^{4+} becomes regular and in the ratio of one aluminum ion for every three silicon ions, mica minerals are again produced. The dark colored mica of this group is *biotite*, and the white, purer mica is *phlogopite*. Both of these micas are found in small to trace quantities in many clays and shales used by our industry.

Last in the classification is an important group of minerals which we call *chlorites*. There are many mineral names in this group, but they all are part of substitutional solid solutions of a basic structure; therefore, each mineral has an arbitrary range of compositions. (This is somewhat analogous to the feldspar minerals in the plagioclase series.) The chlorite minerals have a mixed-layer arrangement in a regular repetition instead of the stacking of the same layers as has been heretofore encountered. The crystals are built up by a regular intermixing of biotite-like layers and brucite layers. First one, then the other are stacked together to form a mica-like mineral [8].

Some of the more common names of chlorite minerals are listed in Table 2, but only a general formula is given. They are listed here in the order of decreasing numbers of magnesium ions in the octahedral positions of the biotite-like layer and in increasing amounts of aluminum ions in these positions. Ferrous ions can substitute for magnesium ions in the biotite-like layer, and ferric ions can replace aluminum in the tetrahedral sheet. These iron substitutions can be only partial, and in the octahedral sheets of the biotite-like layers, the amount of substitution does not exceed one-half of the trioctahedral sites. The ferric substitution is probably even more limited [9]. The general formula reflects the introduction of the brucite layer by having more magnesium and hydroxyls than in the simple trioctohedral minerals.

The chlorite minerals are commonly found in red-firing clays and shales used by the industry, sometimes in major quantities. Those mica-like minerals have important effects on the firing of structural clay products and on the properties of the final product.

Now the difficulties and imperfections in this disilicate classification can be appreciated. After the 2-layer category, the substitutions in the octahedral sheets cause one to wonder whether the minerals are di- or trioctahedral. The decision was made here to classify the minerals as dioctahedral if their octahedral sheets were commonly more dioctahedral than trioctahedral, and similarly for the trioctahedral listing. This does not rule out the more rare possibility that a mineral like illite or even chlorite may be found that is more like its opposite classification. Other classifications of the clay minerals have been made by Mackenzie [10] and Lazarenko [11], and they may be consulted for additional thought on this subject.

2.3. Essential Minerals

Every ceramist knows that a classical whiteware body is prepared with approximately 50% clay, 25% feldspar, and 25% flint. The primary reason for the clay is to provide plasticity and, consequently, workability to form useful shapes. In this plastic state the feldspar and flint act as fillers to control shrinkage and warpage of the pieces formed. On firing, the feldspar becomes a flux, compensating for the refractory clays and inert quartz. The "flint," which is really pulverized quartz, is an inert filler throughout the process of making most whitewares, but it does act to provide strength to the dried and fired ware.

The criteria for a typical clay body is, then, clay for plasticity, a fluxing material, and a filler. These requirements still hold true for structural clay products. For the most part, these ingredients occur naturally in the clays and shales used in this industry. When they do not occur in the right proportions in a natural deposit, either modifications must be made by the addition of other appropriate substances or the material must be discarded as unfit for the purpose intended. In the floor and wall tile industry, some of the bodies are prepared by blending the necessary ingredients for the development of special properties in much the same way as for porcelain.

Unlike a whiteware body which contains refractory kaolinitic clays, the red-firing clays for structural clay products are predominately illite. Because of the presence of potassium and magnesium in the formula of this mineral, it does not require an additional flux. In this sense, the clay mineral contains its own flux. Illitic clays do require an inert filler, however, and this is usually fine-grained quartz which occurs naturally in most such clays and shales. The essential minerals for red-firing structural products are, therefore, illite and quartz. These two minerals are all that are necessary to produce high quality ware; however, other minerals are usually present which may or may not modify the product significantly.

The fireclays, used for buff-firing structural clay products, contain kaolinite as the major clay mineral; however, they also contain smaller amounts of illite. Both clays contribute to the development of plasticity, but illite is also the primary flux on firing. Due to the higher temperatures employed for maturing fireclay bodies, any feldspar that may be present will also have a fluxing action. Quartz is again a necessary ingredient as a filler, and it is usually present in the natural deposits. The essential minerals, then, for a fireclay product are kaolinite, illite, and quartz. In the right proportions this combination produces excellent products. As is the case for most naturally occurring clays, fireclays are apt to contain other minerals that may alter the process or the product.

We have stated that a typical whiteware body contains approximately 50% clay; however, the clay content necessary for the type of product and the method of forming may vary somewhat from this value in structural clay products. The clay-mineral content of the materials used for the extrusion of bricks may be as low as 35%, but this is not to say that a content closer to 50% would not be better. For the extrusion of structural tiles and small pipes, a more desirable minimum in clay content would be about 40%. This would also hold true for the soft-mud brick process. Raw materials for large pipes need to develop great plastic strength; the requirement here is more nearly the typical 50% clay. In all clay products it is undesirable to have much more than 50% clay because of the excessive shrinkage that results.

2.4. Nonessential Minerals

Most natural clay and shale deposits contain minerals not essential for the production of good products. Some of these minerals are not harmful, and some can be beneficial; however, others are detrimental or at least worrisome. The more common nonessential minerals which have been found in raw materials used for structural clay products are listed in Table 3, where they are subdivided into types which reflect both structure and chemistry.

The disilicates, other than the essential clays, are listed first in Table 3. Montmorillonite is a clay mineral that is found in amounts up to about 5% in some useful clay deposits. Due to the unusual behavior of this mineral in the development of plasticity and in drying, materials containing even one or two percent should be processed in specific ways to avoid difficulty. In some factories bentonite is added in small amounts when the plasticity would otherwise be too poor. In such cases, this mineral would become an essential ingredient, but this is by no means common practice. Chlorite minerals occur in many crude clays and shales, sometimes in amounts up to 20%. This mineral is not a clay; it does not contribute to plasticity, but its high magnesium content tends to promote different phases in the firing reaction. The rest of the disilicate micas and mica-like minerals usually do not occur in quantities large enough to affect the products or processes significantly. All of the disilicates listed here except pyrophyllite act as fluxes at usual firing temperatures. Pyrophyllite can be mixed as a relatively pure mineral, and in this state it is an essential raw material for some floor and wall tile bodies. *Sericite*, listed with muscovite, is a type of muscovite presumed to be formed in the earth by weathering of other rocks rather than crystallizing from melts as is the case with muscovite. Finely ground muscovite or sericite is sometimes added to other clay materials in structural clay products factories to act as a flux.

Table 3. *Nonessential Minerals Commonly Found in Clays and Shales for Structural Clay Products*

A. Disilicates	zircon
montmorillonite	garnets
chlorites	
muscovite and sericite	F. Oxides
biotite	rutile
vermiculite	anatase
pyrophyllite	hematite
	limonite
B. Network Silicates	ilmenite
microcline	
orthoclase	G. Carbonates
albite	calcite
oligoclase	dolomite
andesine	siderite
labradorite	
tourmalines	H. Sulfate
	gypsum
C. Amphibole Silicates	barite
tremolite	
hornblende	I. Sulfides
	marcasite
D. Metasilicates	pyrite
diopside	pyrrhotite
augite	
	J. Elements
E. Orthosilicates	carbon (lignite or coal)

The network silicates are usually found in small to trace quantities in most clays and shales. *Microcline* and *orthoclase* are potassium feldspars, and from *albite* to *labradorite* in the table are the soda-lime feldspars of the *plagioclase series*. These feldspathic minerals act as inert fillers throughout the process of making red-firing products, but they can act as fluxes in the higher temperature, fireclay materials.

The *amphiboles*, meta-, and orthosilicates, are only found in small to trace quantities in materials for structural clay products, and their effects are usually insignificant.

Of the oxides listed, only *hematite* and *limonite* are of concern to these clay products. These iron oxides greatly affect color control and their high temperature chemistry must be thoroughly understood. The titanium dioxides, *rutile* and *anatase*, are widely distributed in the surface of the earth's crust; therefore, one could expect to find a percent or so of these minerals in most clays and shales. At this level of concentration, they have little effect on products formed from them.

The *carbonates* of calcium and magnesium are highly reactive with the clay minerals, and their presence leads the high temperature reactions into entirely different directions. This is not necessarily harmful since the reaction products are often beneficial, but raw materials containing these impurities are often avoided. Excessive amounts of *calcite* and *dolomite* in clay materials will result in a completely unstable product. *Siderite*, ferrous carbonate, is often found in clays as small to large nodules, and it is probably derived through the decomposition of pyrite (FeS_2). Because these nodules are harder to pulverize than the rest of the clay, the larger particles resulting cause iron spots on the products. In the structural clay products industry, this behavior of siderite is usually undesirable.

The sulfates of calcium and barium are quite undesirable because gypsum is somewhat soluble in the water used for achieving plasticity, and both sulfates are a source of highly reactive sulfurous gases. The presence of these gases is even more detrimental to the product than they are to the environment.

Under sulfides, all three minerals listed are iron sulfides. They are found in most fireclays and many dark-colored clays and shales where organic carbon has prevented their natural oxidation. To say the least, these minerals are troublesome, and we could do without them; however, they can be handled technically in any firing process which allows otherwise good raw materials to be used satisfactorily.

Carbonaceous matter is present in many clays and shales, and it must be burned out at a particular stage of the firing process to avoid serious difficulties. This can be accomplished, but it may add considerable time to the firing schedule. This is a problem that is successfully handled by the structural clay products industry.

2.5. Typical Compositions

It would be quite impossible to present all of the mineralogical compositions used by the various plants, since every natural deposit is probably different in varying degrees; however, there are always certain basic similarities, such as, the contents of clay, filler, and flux or reactive ingredients. There are also basic similarities in red-firing clays, shales of various geological ages, and fireclays.

Some typical compositions of clay materials used by the structural clay products industry in the United States are given in Table 4, where the analyses are grouped under red-firing materials and those that fire to lighter colors.

The red-firing materials include recent clay sediments and shales of older origin. Note the large amount of chlorite in the hard shale. This is typical and the relationship will be discussed again in the next chapter. Note also the relation between the presence of microcline and kaolinite. This arises from the fact that on weathering the potassium feldspars degrade to kaolinite and mica. Another combination worthy of note is the occurrence of montmorillonite in black clays with organic matter present. This is probably a swamp-land clay where vegetation grew and drainage was poor. These conditions tend to promote the formation of montmorillonite from other silicate minerals. The best red-firing material presented here, producing the fewest problems in manufacturing, is the brown clay. This is, no doubt, a flood-plain clay in some river valley.

The buff to white-firing clay materials listed show three ways by which light colors are produced. Fireclays fire to buff shades because kaolinite is the essential clay

Table 4. *Typical Mineral Compositions for Structural Clay Products (Percents)*
A. Red-Firing Clays and Shales

Mineral	Hard, Blue Shale	Brown Clay	Red Shale	Black Clay	Soft, Gray Clay	Blue Clay
Quartz	40	49	46	39	45	48
Illite	30	40	37	29	20	40
Kaolinite				10	20	
Montmorillonite				2		
Chlorite	14		5			5
Sericite	6	5		5	4	
Microcline				10	8	1.0
Oligoclase	5	4	4			1.0
Calcite	3		0.5			2
Gypsum			0.5			0.5
Pyrite	0.3				0.5	0.5
Carbon	0.7			4	1.5	1.0
Rutile	1.0	1.0	1.0	1.0	1.0	1.0
Hematite			6			
Limonite		1.0				

B. Buff to White-Firing Clays and Shales

Mineral	Fireclay	Gray Shale	White Clay	Blue Clay
Quartz	49	52	40	38
Kaolinite	35		40	
Illite	10	30		40
Chlorite				7
Muscovite			10	
Microcline	2		10	2
Calcite		15		10
Gypsum		2		
Pyrite	1.0			0.5
Carbon	2			1.5
Rutile	1.0	1.0		1.0

mineral and the total iron oxide content is relatively low, perhaps 3 to 4 percent. The gray shale and the blue clay are buff firing because of the presence of calcite ($CaCO_3$) in significant amounts. The white clay is a typical residual kaolin, and it fires white simply due to the absence of coloring oxides. The chemical reasons for the development of light colors with these assemblages of minerals will be presented in detail in Chapter 6.

Not included in Table 4 are the prepared compositions used in the production of floor and wall tiles. Some of these tiles are made from clays and shales with compositions similar to those listed or combinations of them; however, other floor and wall tile bodies are prepared from purer materials blended like a porcelain body to produce the special properties required of these products. Some of these compositions may contain pyrophyllite, talc, or wollastonite in addition to the proper blend of kaolin and ball clays and a feldspathic flux [12].

Perhaps it has been noted by now that no oxide chemical analyses have been used to characterize clays or body compositions for structural clay products. This is because the mineralogical analyses are more meaningful in this industry than the typical chemical analysis. In the first place, the kinds and quantities of the clay minerals present determine the plasticity and workability of the materials. They also need to be in proper balance with the other minerals to provide formability without excessive shrinkage which can lead to warping and cracking of the pieces formed. The paths of the high temperature chemical reactions which take place on firing are entirely dependent on the minerals present and have little relation to the percentages of oxides in the composite raw material. The mineralogical analyses also show the reactive ingredients and fluxes that are present which, in turn, suggest the temperatures at which these minerals become active. There is no way these bits of important information can be read from a chemical analysis. On the other hand, chemical analyses and determinations can be useful in certain very specific problems. When these are considered, chemical analyses will be presented.

References

1. Pauling, L.: The structure of micas and related minerals. Proc. Nat. Acad. Sci., U.S. **16**, 123 (1930).
2. Brindley, G. W., and K. Robinson: Randomness in the structures of kaolinitic clay minerals. Trans. Faraday Soc. **42B**, 198–204 (1946).
3. Weir, A. H., and R. Greene-Kelly: Beidellite. Am. Min. **47**, 137–45 (1962).
4. Phillips, G. C., Jr.: Behavior of Illite on Heating. M.S. Thesis, N.Y. State College of Ceramics, June, 1964.
5. Yoder, H. S., Jr.: Experimental Studies on Micas: a Synthesis, in Clays and Clay Minerals, E. Ingerson, ed., pp. 42–60. New York: Pergamon Press. 1959.
6. Brindley, G. W., ed., X-ray Identification and Crystal Structures of Clay Minerals, pp. 199–212. London: The Mineralogical Soc. 1951.
7. Gruner, J. W.: The structures of vermiculite and their collapse by dehydration. Am. Min. **19**, 557–75 (1934).
8. Brown, B. E., and S. W. Bailey: Chlorite polytypism, I, regular and semirandom one-layer structures. Am. Min. **47**, 819–50 (1962).

9. Nelson, B. W., and R. Roy: Synthesis of the chlorites and their structural and chemical constitution. Am. Min. **43**, 707—25 (1958).
10. Mackenzie, R. C.: Classification and nomenclature of clay minerals. Clay Min. Bull. **4**, 52—66 (1959).
11. Lazarenko, E. K.: Nomenclature and classification of clay minerals. Clay Min. Bull. **4**, 67—68 (1959).
12. Emrich, E. W.: History and development of ceramic wall tile bodies in the United States. Am. Ceram. Soc. Bull. **52**, 687—88 (1973).

3. Raw Materials and Processing

3.1. Mining

3.1.1. Exploration for Raw Materials

Now that we know what to look for in materials for structural clay products, the next step is to locate proper deposits and to win them in the best possible way. A knowledge of the science of geology is extremely helpful here because the application of geological principles changes the search from a needle-in-a haystack approach to predicting where satisfactory materials might be found. Geological maps are available from state and federal governments for large areas of the United States. Such maps may be bedrock maps which are especially helpful in locating shale deposits or they may be soil maps that indicate the presence of clay deposits. Where these maps are available, they should be used as a guide for exploration, but if they are not available the next best aids are topographic maps which are available for the whole country from the federal government. Some of the large structural clay products companies have a geologist who is assigned to find deposits, maintain adequate reserves, and oversee the mining operations. Other companies should hire a consulting geologist who is familar with the area to assist in exploration.

At the outset of exploration for new materials, the observation of certain topographical features of the landscape are helpful. Stream, highway, and railroad cuts often provide excellent views of the materials normally hidden beneath the surface. Although a good deposit may not always be used at the place where it is first seen, the geologic and topographic maps may be used to trace the deposit to a more favorable mining location. In the search for clay deposits, areas of poor drainage often indicate that an impervious clay layer lies below the surface. Flood plains of larger streams and rivers may contain sedimentary clay deposits.

The various kinds of important clay deposits provide more specific information for uncovering new deposits for specific purposes. The residual white kaolins, used only sparingly by this industry but important for the production of white face bricks, are found in areas where acidic, feldspathic, igneous rocks may have been weathered to kaolinite. To find such deposits which have escaped glaciation, one should look in narrow deep valleys transverse to the direction of glaciation. Residual deposits may be buried under other glacial debris. Obviously, unglaciated areas are apt to be more productive in residual kaolins as is the case in our southern states [1].

Most fireclay deposits, but not all, are associated with the coal deposits of the Pennsylvanian period. A good example of a deposit of this kind will be presented later in this chapter as Fig. 20. Some of the best fireclay deposits are found under beds of coal, and they are often mined after the removal of the coal in both surface and underground mines. Scattered here and there are sedimentary fireclay deposits not associated with coal, like the clays of the Cretaceous period on Long Island in New York State. These beds are often lenticular and show considerable lateral variation. These deposits may have been subjected to erosion before burial, and as a consequence, fireclays are completely absent in some places within a deposit. For these reasons reserves should be proven by closely spaced drillings before decisions on their use are made. With these clays color is no guide to quality, because various amounts of organic matter may alter their appearance without changing their usefulness to the industry [1].

In order to be able to speak the language of the geologist and to orient ourselves in the nature of things in the earth's outer crust, Table 5 has been included here showing the geologic time divisions after which many deposits are described. In this table the Cenozoic is the most recent era while the Archeozoic is the oldest [2]. In exploration of an area where the exposed shales are of the Devonian period, as they are over most of New York State, one could not expect to find the Pennsylvanian fireclays, since they would have to lie above the Devonian shales.

Table 5. *Geologic Time Divisions*

Era	Period
Cenozoic (65* MY-Present)	Quaternary (2 MY-Present) Holocene (Recent) Pleistocene (The Ice Age) Tertiary (65–2 MY)
Mesozoic (225–65 MY)	Cretaceous (136–65 MY) Jurassic (190–136 MY) Triassic (225–190 MY)
Paleozoic (570–225 MY)	Permian (280–225 MY) Pennsylvanian (325–280 MY) Mississippian (345–325 MY) Devonian (395–345 MY) Silurian (435–395 MY) Ordovician (500–435 MY) Cambrian (570–500 MY)
Proterozoic (2100–570 MY) Archeozoic (4500–2100 MY)	

* Million years—MY

Recent sediments provide clay deposits which vary both as to mineralogy and size. Some lake or marine clays show close relationships with the surrounding country rocks. For example, some glacial lake clays, in areas where shales are outcropping, have mineralogy identical with that of the shales from which they weathered. The river alluvial clays are usually extensive parallel to the river course but limited in the perpendicular direction. Some recent marine sedimentary clays are remarkably uniform over large areas [1].

In searching for shale deposits, we are likely to come across three types of shales that are visibly alike but mineralogically different, and only one type is suitable as a source of clay for structural clay products. These three types are argillaceous, siliceous, and calcareous. The argillaceous shales contain clay and were probably formed from normal clay sediments. Siliceous shales were formed from silty deposits such as quicksand, and they contain very little clay. These shales can only be used as inert (nonplastic) fillers added to a clay to lower excessive shrinkage or to produce an esthetic texture on bricks. Calcareous shales contain a continuously variable amount of calcite or dolomite, the basic ingredients of limestones. These shales grade from argillaceous shale to limestone. They might be called transitional deposits. Calcareous shales with low percentages of these carbonates can be used in this industry with some special attention, but those with greater amounts are completely unsuitable, as we shall discover later in Chapt. 6.

Another variation in argillaceous shales must also be appreciated in the selection of a satisfactory deposit. This is the gradation of these shales into slates. A good argillaceous shale or a regular slate are not difficult to recognize, but the transitional shales require mineralogical examination to see how far advanced they are toward slate. In this transition the disilicate mineralogy gradually changes from argillaceous shales containing illite clay to slates which are essentially chlorite and muscovite mica or sericite. Slates are impossible to use as raw materials for structural clay products except as filler because they contain no clay for the development of plasticity; therefore, as shales grade toward slates they become less workable due to lower and lower plasticity. Some materials which would be classified as slaty shales are being used in this industry with great difficulty. Such materials should have been avoided.

3.1.2. Testing and Evaluation of Deposits

Clay and shale deposits that look good in the field on the basis of the criteria just set forth should be sampled and tested to determine if, in fact, the material has the desired properties. How are they to be tested? What are the most important characteristics to observe? A recommended procedure is indicated by the sample data sheet presented as Table 6. This data sheet can be used to record pertinent information from the field, on the plastic properties, mineralogy, and firing behavior. A plastograph such as The Brabender Plastograph is very useful for studying the plastic behavior of clay materials in the laboratory. It records the consistency of the sample as water is slowly added to the dry material until it reaches a slip or thick paint consistency. The linear drying shrinkage is measured on test specimens made in the plastic state with the amount of water adjusted to give the maximum consistency. The quick test with cold and hot hydrochloric acid tells immediately whether or not calcite and dolomite are present. Calcite effervesces with the evolution of carbon dioxide in cold acid, but dolomite reacts only in hot acid. Differential thermal

Table 6. *Evaluation Sheet for Clay and Shale Raw Materials*

Field Data

Sample No:_____ Date:_____ Collected by:_____
Location of deposit:
Type and visible appearance of deposit:

Plastic Behavior

Sample screened through_____mesh sieve.
Interpretation of plastograph (attach chart):
Water of plasticity in %:_____
Linear drying shrinkage in %:_____

Mineralogy

Reaction with HCL: (cold)_____ (hot)_____
Interpretation of DTA (attach chart):
Interpretation of TGA (attach chart):
% retained on 270 mesh sieve (wet screening):_____
Mineralogical analysis (semiquantitative):

Firing Behavior
Results from thermal-gradient furnace

| *Temperature* | *Color* | *Scum* | *Shrinkage* | *Absorption* |

Evaluation Comments

---

analysis (DTA) and thermogravimetric analysis (TGA) combined show semiquantitatively the presence of carbonaceous matter, pyrite, calcite, dolomite, and may give clues to the type of clay minerals present. The wet sieve analysis gives an indication of the amount of fine particles which may be clay, and the mineralogical analysis is best done by X-ray diffraction. Preliminary data on firing behavior can be done easily and quickly in a thermal-gradient furnace.

At any point in this scheme of preliminary evaluation, the material may be rejected as unsuitable for the purpose intended. For example, if test specimens cannot be made satisfactorily because of low plasticity or if the shrinkage is too high, the material may be classified as unsatisfactory. If the mineralogical studies show large amounts of dolomite or pyrite, this may be sufficient grounds for rejection. In some cases every property will seem to be satisfactory, but the firing color will rule out the sample completely.

If the deposit of interest turns out to be satisfactory from the first preliminary examination, it must be sampled at frequent intervals as with a 50-foot (15-meter) grid. The dimensions of the grid for sampling will depend to some extent on the type of deposit being explored. Less frequent sampling may be needed if the geologic expectations favor uniformity, as with a shale or a marine clay; however, with residual, glacial lake, or recent flood plain clay deposits, the grid should prescribe more frequent sampling sites. Samples should be taken at all points on the grid by core or auger drilling in such a way as to recover as much material as possible. All of this is done to prove the uniformity of the entire property and to calculate the tonnage available. In some cases where an exposed and weathered surface was sampled first, drastic changes will occur when completely unweathered material is examined. The evaluation of the samples so taken can be made as suggested for the preliminary sample.

When the drilled samples continue to show a raw material satisfactory both in quality and quantity, a representative sample of a few hundred pounds should be mined, dried, ground and quartered down to about 25 pounds for more complete testing related to factory production. Complete mineralogical and chemical analyses need to be made. From the preliminary analyses, it may be desirable for chemical analysis to include determinations for sulfide sulfur, soluble sulfates, free carbon, and carbonate carbon in addition to the usual silicate analysis. More detailed testing on drying behavior is necessary, emphasizing the time and conditions for optimum drying rate without cracking. Many test specimens should be made for use in firing tests, using the method of forming anticipated for production. An antiscumming additive should be included in the batch, if necessary. The fired properties of the material should be examined after firing with the same fuel, atmosphere, and time expected in the factory. (The firings should be made to 5 or 6 temperatures at 50°F (25°C) intervals above and below the optimum maturing temperature.) This examination should be based on color, shrinkage, absorption, saturation coefficient, and efflorescence tendencies. These final results can be used to design new factory equipment for the processing of this particular material or to adjust an existing facility to accommodate the new material successfully.

3.1.3. Mining Procedures

In some locations good clay deposits are below the water table or in areas of slow drainage, and in these cases they are often very wet and sticky—sometimes containing

Fig. 18. Mining wet clay with a dragline

up to 40% water. Such clay pits must be pumped continuously to keep the water out. Drainage ditches dug below the lower level of the mining operation can help in removing the excessive moisture from the clay deposit. In some cases these drainage ditches must be pumped out to lower elevations to effect the continuous removal of water. Where deposits remain quite damp and sticky, it is impossible to mine the clay with shovels and scrapers, and draglines are used to win the clay. An operation of this kind is shown in Fig. 18.

Drier clay deposits containing moisture from 5 to 20% and soft shales can be mined with large earth scrapers or power shovels as seen in Fig. 19. Harder shales must be blasted before mining with shovels or front-end loaders.

Fig. 19. Mining soft shale with a power shovel

A typical dry fireclay deposit is shown in Fig. 20. The readily recognizable sequence of layering from top to bottom is shale, limestone, coal, and fireclay. In this situation the shale, limestone, and coal must be stripped in order to recover uncontaminated fireclay, the useful material in this case.

In all of the preceding figures distinct layering of these sedimentary deposits is visible. Each layer represents a different deposit with respect to time; so mineralogical variations may occur from layer to layer. To mine this type of sedimentary deposit for greatest uniformity, it is necessary to remove the material perpendicular to the bedding planes no matter what the angle of bedding may be in the deposit.

A good practice in mining clays and shales to effect greater uniformity is to stockpile the mined material. After first removing the material perpendicular to the bedding, it is laid down again in thin horizontal layers with scrapers or bulldozers. This stockpile is built, with compaction, to a height convenient for remining with power shovel or front-end loader as pictured in Fig. 21. The location of the stockpile should be as close to the plant as possible in an area where good roads can be maintained

Fig. 20. Typical fireclay deposit underlying coal measures

in poor weather. Another advantage in stockpiling is that the initial mining can easily be completed in the favorable weather months of the year. In the northern latitudes stockpiles may be covered with straw or sheet plastic to keep the stockpile from

Fig. 21. Remining a stockpile with a front-end loader

freezing and for easier removal of snow accumulation. Somewhat smaller stockpiles can be built under the protection of a shed to keep the material more uniformly dry.

This procedure can also be used for wet clay as a means of air drying it to a more satisfactory water content for the forming operation.

A caution should be interjected here concerning a danger in stockpiling that must be avoided. If pyrite is present in the raw material and especially if calcite or dolomite is also present, stockpiles cannot be held longer than two months either inside or outside. This is because pyrite oxidizes to iron oxide and sulfuric acid upon exposure to air. This surely takes place after mining or even blasting a deposit open. The sulfuric acid immediately attacks the other minerals present to produce sulfates which can be troublesome and can make the whole stockpile completely useless.

3.1.4. Mining Pollution Controls

It is reasonable to expect that local, state, and federal governments will continue to insist on the reclamation of mined-out areas. For the structural clay products industry, this has proven to be an advantage in good community relations and economically profitable in some places. It is not difficult, if planned ahead, to convert mining areas into real estate which has considerably greater value after mining than before. The land can be reclaimed for recreational areas including ponds, grazing and crop lands for agricultural purposes, or industrial and residental building sites. Some structural clay products companies have built new plants on reclaimed land.

The general procedures to be followed by the industry for the control of mining pollution can be taken from state laws already in effect which seem to be working satisfactorily. Most likely, governments are going to require permits for mining operations which will be contingent upon detailed plans for reclamation once an area is mined out. The plan will require procedures for handling surface water drainage to avoid pollution of streams, rivers, and lakes.

Such long range planning is useful because it makes the job of restoration easier in the end. For example, the stockpiling of top soil, if available, is an economical move, and the prudent handling of overburden wastes for backfilling is advantageous. Reclamation plans may also dictate the best plan for handling surface water drainage while the mine is in operation.

Progress reports will probably be required and time limits for completed reclamation will be set. This will require careful production planning which the operators will find helpful. Probably bonds will have to be posted to cover the costs of reclamation, and the operations will be subject to inspection at any time to ascertain if the plans approved are being carried out.

3.2. Raw Material Processing

Wet clay must be dried to a water content below that required for forming, since mixers, such as pug mills, are designed to mix while water is added, even in small amounts. They do not work well when attempting to temper a too-wet material with a dry material to bring about the proper consistency. Some factories use rotary, fuel-fired dryers or stockpile-and-turnover procedures for air drying.

Damp clays that cannot be screened are put through smooth rolls to break up big lumps, to crush pebbles, and to start the mixing process. The output of the rolls will be irregular slabs of clay about one inch (2.5 cm) thick, and these will go directly to a pug mill where additives are made and water is added to produce the desired consistency. In the past such damp clays were used exclusively for soft-mud bricks molded in sand-coated boxes, but in recent times extruded products are also made, even though it may be necessary to add a relatively dry shale, clay, or calcined raw material to control the moisture content and plasticity.

A different process is used for preparing dry clays and shales. The equipment is similar for both materials, but there are differences in the details of the machinery when going from dry clay to a hard shale. The problem is to grind the clay-containing materials to fine particle size, and this requires several separate operations as the feed becomes smaller and smaller.

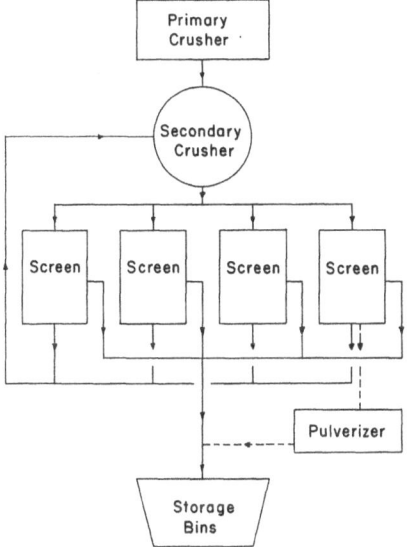

Fig. 22. Flow diagram for processing dry clays and shales previous to forming

Fig. 22 presents a typical flow diagram for such a comminution process. The fixed-path primary crusher is in series with the secondary crusher; so they both must have the same capacity for continuous operation. Sometimes a surge bin is placed between these operations so that they do not have to have equal capacities. In other cases there may be two or more secondary crushers receiving the feed from the primary crusher. The secondary crushing operation and the screening is a closed-circuit system. The material being ground passes continuously through the machines to the screens where the fine particles are removed for production, and the coarse particles are cycled back to the secondary crusher.

The pulverizer flow pattern is designated by dashed lines to indicate an optional path. For some shales part of the material not passing the screens is sometimes fed into a pulverizer where the output is all of subscreen size. This increases the amount of fine particles in the final distribution which, in turn, increases plasticity in production.

The primary crusher is designed to take the material as mined, which may be in lumps or pieces up to one foot (0.3 m) in diameter. The machinery to break down such large pieces needs to be especially strong and powerful. For this purpose jaw, double-roll, or single-roll crushers are used. The roll crushers are not smooth rolls as are used with damp clays. They often have knobs or teeth on the rolls to pull the odd-shaped chunks into the roll mechanism and to assist in breaking them. The double-roll crushers feed between the rolls which are set apart to control the size of pieces passing, and at least one or both rolls are strongly spring loaded so that they will not become jammed if a particularly tough piece is forced into the mechanism. A single-roll crusher crushes between the roll and a perforated plate, as is shown in Fig. 23.

Fig. 23. Single-roll crusher receiving a charge of shale as mined

A dry clay crusher is more apt to have teeth instead of knobs on the rolls. The distribution of particle sizes after primary crushing is from about 2-inch size (5 cm) down to a fine powder; however, the amount of fines will be extremely small, since this machinery is not designed to produce small particles.

The product from the primary crusher feeds the secondary crusher which may be a dry pan, rim-discharge grinder, hammer mill, or chain mill. All of these machines are designed to grind the 2-inch material down to the size of the screen openings and finer. The hammer and chain mills use rotating and swinging objects to beat the material against a perforated plate to reduce the particle size. The dry pans and rim-discharge grinders use two large, rotating and turning, steel wheels to crush the charge against a circular bottom plate. The dry pan has a perforated bottom plate for the particles to pass when they are reduced to about 1/2 inch (1 cm) in diameter. The rim-discharge grinder discharges the ground material over the side of an adjustable-height pan resting on a rotating base plate. A pair of rim-discharge grinders is shown

Fig. 24. Two rim-discharge grinders with a fireclay feed

in Fig. 24. They are receiving fireclay from a surge bin which holds the output from a single primary crusher.

All of these secondary crushers are designed to discharge when only part of the charge is fine enough to pass the screen openings. This is done to prevent an accumulation of material in the crusher which will pass the screens and reduce grinding efficiency. The amount of material ground fine enough to go through the screens before discharge varies from 25% to 75% depending on the material being ground and the size of the openings in the screen. The finer screens send more oversize back to the crusher. The feed rate to these crushers must be constant; therefore, the feed from the primary crusher must be regulated to just compensate for the amount of material passing the screens. This is usually done by monitoring the electric power required by the secondary crusher. As the level of material builds up in the crusher, more power is required for operating the machine. An increase in amperage to the motor cuts off the feed from the primary crusher. Conversely, when the amperage falls to a preset value the feeder puts more primary material into the crusher [3, 4].

Each secondary crusher requires a set of screens to remove the material ground fine enough for production requirements. All screens may have the same size openings or they may be different in order to obtain the desired particle-size distribution. In any case sufficient screening capacity must be installed to receive the continuous output from the secondary crusher. As can be seen in Fig. 22, the oversize from the screens is returned to the secondary crusher for continued grinding while the material passing is delivered to bins in order to keep a constant supply for production.

Clay materials, as used in the structural clay products industry, require definite screen characteristics. The screens are inclined, so that the material will flow continuously over the openings. This inclination is fixed, together with the feed rate.

The feed to each screen must be uniform across the screen and the thickness of the bed held constant as it falls on the screen. A bank of screens in operation is shown in Fig. 25. The flow of feed down the screen is from left to right. The electrical vibrating

Fig. 25. Bank of screens in a structural clay products plant

mechanisms are located on the bridge over each screen, and the electrical heating mechanisms·are visible on the lower left sides. The capacity of these screens is more related to width than to length. The efficiency improves with the length as the load lightens, and fine particles are more free to move and find an opening.

Because of the flat particle shapes, even though the ground particles are aggregates of much smaller particles, oblong meshes are often used as screens with the width of the slot being the factor determining size. Since clay materials tend to block or "blind" the screens, the long axes of the meshes are set in the direction of flow of the feed. The minimum size of openings for this type of dry screening is 20 mesh (0.033 inch

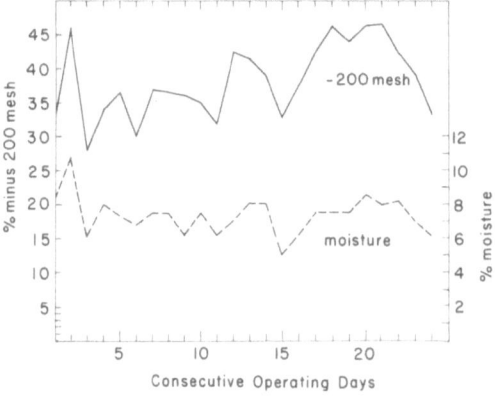

Fig. 26. Fines produced from shale with variable moisture from day to day

or 0.841 mm). Below this size, the screening efficiency falls so low that the screen capacity will not keep up with production requirements [3, 4]. Some structural clay products factories use screens up to 3/8 inch (9.51 mm); however, it is quite usual to have screens of 8 mesh (0.094 inch or 2.38 mm).

In this industry it is uneconomical to dry the raw materials to a constant state before grinding and screening; therefore, variations in moisture, due to the vagaries of the weather and the hygroscopic nature of the clay minerals, cause efficiency to vary in grinding and screening. This not only requires enough screening capacity for the more adverse conditions, but the particle-size distribution in the product is altered according to the moisture content. Such fluctuations are shown on Fig. 26 where the fine particles (−200 mesh) and moisture contents were monitored each day for 24 operating days in a plant using shale as a raw material. The amount of fines increased with moisture content because the damp particles tended to aggregate together and did not pass the screen as they should. In this situation many fine particles were cycled back to the crusher for further comminution.

These variations in the particle-size distribution are, for the most part, absorbed in production without great difficulty; however, there are times and places where these moisture variations cause such disruptive effects as off-grade ware.

3.3. Particle-Size Distribution

The particle-size distribution is more important to consider in the preparation of materials for structural clay products than the size openings of the screens or the tonnage output through the screens. In these products there are three important ranges of particle sizes, each with its own specific function.

First, there is the *texture fraction* of the total distribution, and this is from 3/8 inch to 14 mesh (9.5 to 1.2 mm). The only technical reason for particles of this size is for esthetic purposes in the development of surface textures in the forming processes. Sometimes the distribution may contain particles up to 8-mesh size (2.4 mm) for economic reasons only, but when this is done, the green, dry, and fired strengths are sacrificed. However, it can be possible to take advantage of this economy whenever there is more strength than is necessary to achieve good products with no production losses.

Second, the particle sizes from 14 mesh to 270 mesh (1168 to 53 μ) constitute the *filler fraction*. The function here is to control excessive shrinkage, cracking, and warping of the products formed in the plastic state.

Third, the *plastic fraction* is that portion of the distribution which has particle sizes below 270 mesh (53 μ). As the name implies, this fraction produces the necessary plasticity for forming and strength for handling the green ware. It is this fraction which contains the effective clay particles that are solely responsible for plasticity. It must be borne in mind that not all of this fraction is necessarily clay, but it is an adequate approximation with otherwise suitable raw materials.

These three particle-size fractions can be the basis for routine quality control in the production system, and this makes frequent complete sieve analyses unnecessary as long as the type of raw materials remains reasonably constant. The amounts generally found in the three fractions are:

0% to 30% in the texture fraction,

20% to 65% in the filler fraction,

35% to 50% in the plastic fraction.

These amounts are quite typical of a ground shale and some clays. Some rather silty clays may have more in the plastic fraction, since many of the fine particles are not clay grains, and act as filler. As in the more precision whiteware bodies, the total nonplastic fraction should be from 50% to 65% and this, of course, includes both the filler and texture fractions. The amount and size of the grains in the texture fraction are quite arbitrary depending on the roughness of the texture desired. Whatever is decided upon in this respect must be held constant to produce consistent products. In order to produce a texture fraction with some clays, particularly recent sediments, it is necessary to blend in a ground shale, rock, or grog (prefired clay). Grog is by far the best material for this purpose, but it is frequently too expensive.

In the production of structural clay products, there are several factors which can affect the particle-size distribution. (1) The blending of different raw materials will provide the distribution of sizes in each to the overall distribution. (2) If screens are used in material processing, the screen cutoff point will make a significant affect. (3) The sequence and types of crushing equipment will impose their own characteristics of grinding action on the distribution. (4) Multimineral raw materials contain particles that exhibit different crushing behaviors; so the mineralogy can affect the particle-size distribution even though it may be in rather insignificant ways. (5) A very important factor in this industry is the moisture content of the raw materials being processed. It is important to determine the particle-size distribution and to be able to relate it to these five factors.

The particle-size distribution must be determined in the laboratory with standard equipment and procedures because production screens do not compare accurately with their laboratory counterpart, and in production the particles tend to be smaller than the designated size of screen opening. Standard sieves with square openings are used in the laboratory to measure particle sizes down to 270 mesh, which is 0.0021 inch or 0.053 millimeter. Procedures for sieve analysis are specified by the manufacturer of standard sieves [5]. Subsieve sizes may be determined by methods employing a pipette, hydrometer, Coulter counter, an electron photomicrographic Zeiss particle-size analyzer, adsorption technique, and others [6, 7].

One must keep in mind just what is being measured with the various techniques. It doesn't require too much imagination to perceive that in a sieve analysis one is measuring an intermediate dimension—not the longest or the shortest. If the analysis is extended beyond the sieve range with another method, the actual thing being measured must be appreciated. Sometimes it may be an equivalent spherical diameter, sometimes particle volume, surface area, or some arbitrary dimension. There is no reason to expect anisodimensional clay particles to give comparable results with different methods; however, satisfactory accommodations can be made among different methods if one understands the differences.

For clay materials a wet sieve analysis gives more appropriate data because the clay aggregates tend to separate into smaller sizes when placed in water. Some of this separation actually takes place in production when water is added to produce plasticity; therefore, perhaps a wet-sieve analysis more nearly describes the sizes as the factory equipment sees them.

In making a sieve analysis the complete $\sqrt{2}$ or 1.414 series of sieves, as provided by the W. S. Tyler Company, should be used. This series gives equally spaced data along a logarithmic particle-size axis. It is based on the Tyler 200-mesh sieve with openings of 0.0029 inch or 0.074 millimeter, and each sieve of the series has openings larger and smaller in multiples of 1.414 [5].

Normally, a particle-size analysis is reported as cumulative percentages either coarser or finer than each size measured. Plots of such data are shown in Fig. 27 for a clay and a shale brick mix. Both are blends of two separately-mined raw materials, and both are used for the production of face bricks. The cumulative curve of a normal, bell-shaped distribution of particles should be a complete S-shaped curve. Note that neither curve shows this characteristic. The large particles of the shale distribution

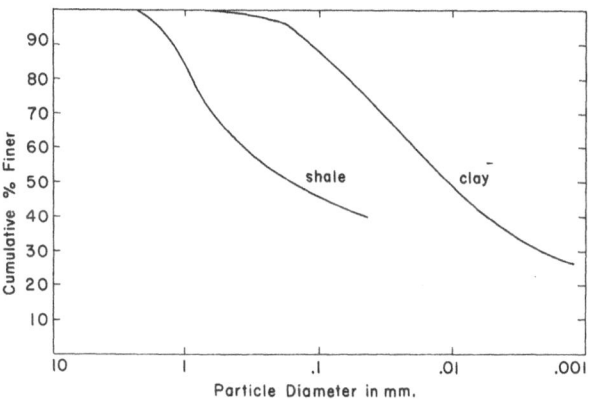

Fig. 27. Cumulative curves of particle sizes in a clay brick mix and a shale blend

were cut off by the screen, and the subsieve sizes were not completely determined. All one can say about this is that about 40% of the shale is below 270 mesh or that there is 40% plastic fraction. On the other hand, the clay shows a more complete distribution towards the large sizes, but the fine-particle end of the curve is still somewhat undetermined, and there is 27% finer than 2 microns. Notice that the curves are not smooth S-shaped curves. The sudden changes in slopes and irregularities mean something about the distribution, but it is difficult to see the distribution from these curves.

One could plot histograms representing the amount of material in the intervals between the particle sizes actually measured, but they are not realistic for sedimentary distributions, and they are not sensitive enough for good visual interpretation. A better way to view and interpret the nature of the distribution of particles according to size is to plot a frequency curve. Such a curve must be derived from the cumulative curve drawn from the experimental data.

Krumbein [8] has shown that sediments give highly skewed frequency curves when the data are applied to the Gaussian normal curve of error, but when the size data are expressed with a logarithmic measuring scale, the curves become much more symmetrical. At the same time, he showed that the introduction of such a logarithmic size scale does not change the characteristics of the Gaussian curve.

Another good reason to transform the size scale from arithmetical to logarithmic is the ability to plot the frequency distribution curve on linear graph paper. Krumbein suggested a logarithmic *phi* scale be used instead of the linear millimeter scale to express particle sizes. The transformation is made by Equation 1,

$$\phi = - \log_2 x \quad \text{or} \quad x = 2^{-\phi} \tag{1}$$

where phi (ϕ) is the new logarithmic scale for particle diameters, and x is diameter expressed in millimeters. A helpful table of equivalents in the expression of particle sizes is given in Table 7.

Table 7. *Particle Size Equivalents*

Mesh*	Inches*	Millimeters	Phi
4	0.185	4.70	-2.25
6	0.131	3.33	-1.75
8	0.093	2.36	-1.25
10	0.065	1.65	-0.75
14	0.046	1.17	-0.25
20	0.0328	0.833	0.25
28	0.0232	0.589	0.75
35	0.0164	0.417	1.25
48	0.0116	0.295	1.75
65	0.0082	0.208	2.25
100	0.0058	0.147	2.75
150	0.0041	0.104	3.25
200	0.0029	0.074	3.75
270	0.0021	0.053	4.25
400	0.0015	0.038	4.75
		0.026	5.25
		0.019	5.75
		0.013	6.25
		$\vdots$	$\vdots$
		0.00082	10.25

* Tyler Standard Sieve Scale

The mathematical procedure for calculating a unique frequency curve from the experimental cumulative curve was worked out by Brotherhood and Griffiths [9]. They found the application of Stirling's Interpolation Formula to be quite satisfactory. This formula is

$$y = y_0 + \frac{x}{1} \cdot \frac{\Delta_0^1 + \Delta_{-1}^1}{2} + \frac{x^2}{2!} \cdot \Delta_{-1}^2 + \frac{x(x^2 - 1)}{3!} \cdot \frac{\Delta_{-1}^3 + \Delta_{-2}^3}{2} + \cdots \tag{2}$$

where y_0 is the central value and (x_i, y_i) is a known series of values in which

$$i = -n, -n + 1, \ldots, -2, -1, 0, 1, 2, \ldots, n - 1, n$$

For application of the Stirling equation to the particle-size distribution, the intervals in x, particle sizes in phi units, are chosen equal. Since Krumbein [10] pointed out that the frequency curve is the derivative of the cumulative curve, the derivative of Sterling's formula is taken after modifying it for equal intervals in x and by assuming that y is very close to y_0. The resulting expression is

$$\frac{dx}{dy}\bigg|_{y=y_0} = \frac{1}{h}\left(\frac{\Delta_0{}^1 + \Delta^1{}_{-1}}{2} - \frac{\Delta^3{}_{-1} + \Delta^3{}_{-2}}{6 \times 2} + \frac{\Delta^5{}_{-2} + \Delta^5{}_{-3}}{30 \times 2}\right) \tag{3}$$

where Δ^1, Δ^3, and Δ^5 are the 1st, 3rd, and 5th differences between consecutive pairs of y values which are the cumulative percentages at each selected size.

Equation 3 gives the slope of the cumulative curve at any point y_0 for the increment h in particle size using the phi scale. The linear plot of these slopes gives the unique frequency distribution curve. In solving the above equation for the chosen intervals, it is unnecessary to go beyond the third differences in percentage values (y_0).

At the beginning of this calculation, it is convenient to plot the cumulative curve of percents on linear graph paper using the phi scale for particle sizes. The interval h can be taken as 1/2 phi as presented in Table 7, or more points can be generated by reading, from the carefully plotted cumulative curve, the percents at 1/4-phi intervals for illustrating more details of the distribution.

The calculation is made fairly simple by putting the phi intervals in a column and recording the corresponding percents from the cumulative curve opposite the phi values. Then the first, second, and third differences between the percents are taken. The third differences are summed in pairs and multiplied by $-1/6$. Next the sums of pairs of the first differences are recorded and the sums of the third differences are added to the sums of the first differences. Lastly, these added values are multiplied by $(1/2)h$ to obtain the frequency of the appropriate phi value. The frequencies are then plotted against phi values to obtain the unique frequency curve for the distribution.

Fig. 28 are unique frequency plots of the same distributions shown in Fig. 27. Quarter-phi intervals were used in these calculations. The plot was made on the phi scale, but equivalents are shown for millimeters and mesh sizes. These distributions are immediately more meaningful visually. The screen cutoff point on the shale distribution is clearly visible on the large-particle-size end of the distribution. The particle size occurring most frequently in this distribution is 0.50 phi (0.707 mm or 24 mesh). Since the shale curve does not show a distinct bimodal character, it appears that the two shales blended had similar grinding characteristics. This shale distribution is typical of that obtained with a dry pan or rim-discharge grinder. Both distributions show considerable unmeasured fines. Note that the clay distribution is definitely bimodal, which indicates the blending of two raw materials with different distributions of particles within themselves. The broader distribution of particles in the unconsolidated clay is typical of these deposits. The sharper distribution of the shale is created by the pulverizing process.

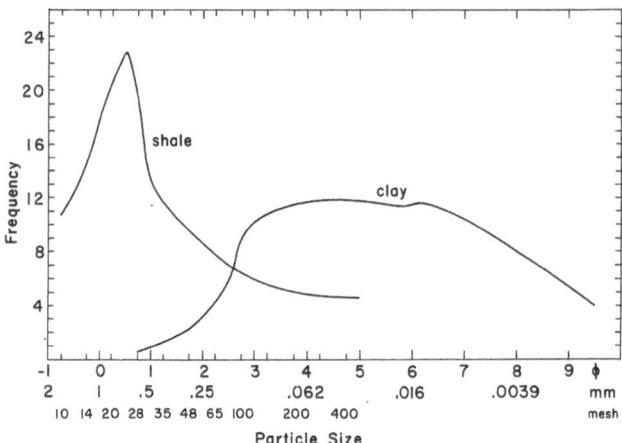

Fig. 28. Unique frequency curves of the particle-size distributions of the previous
figure

Since the unique frequency curve is a derivative of the cumulative curve, the
frequency curve is very sensitive to slight changes in slope of the cumulative curve.
This means that great care should be taken in obtaining the best experimental data
possible and in drawing the continuous curve through these points when constructing
the cumulative curve. A full discussion of the limitations and accuracies of this
method of presentation of data is described by Brotherhood and Griffiths [9]. The
factors, previously listed, that affect the particle-size distribution tend to influence
the unique frequency curve away from a normal distribution. With some knowledge
of the processing, these factors often make themselves visible on the unique frequency curve.

3.4. Dust Pollution Controls

All of the stages of dry mineral processing require dust collection devices for health
safety, worker morale, reduced equipment maintenance, and operating efficiency. At
this time most state governments have laws requiring dust collection and setting
standards for the quality of the air surrounding such operations. Existing plants
must design effective dust collectors around their present equipment, and a dust
collection system must be an integral part of the design of any new plant.

One side effect of the operation of dust collection equipment should be kept in
mind. Most of this equipment is inherently noisy; therefore, a dust pollution problem
could be traded for a noise pollution problem. If this is recognized in the beginning,
the blowers and motors can be isolated from human ears by acoustical insulation,
but this may be nearly impossible after installation.

It is quite possible that the dust collected from the processing of minerals can be
utilized regularly in the production. The dust will become a part of the plastic fraction

if it is added back to the processed material, and one must recognize that the clay mineral content may be higher in the dust than in the material as a whole. Such an addition may be beneficial where the degree of plasticity is consistently low.

3.5. Blending and Additives

Dry ground materials ready for processing are stored in large surge bins that may contain enough material for several days production. The purpose of storage at this point is to prevent production delays when mechanical failures or adverse weather conditions might otherwise cause the entire plant to be shut down. Separate bins are required for each different material or different particle-size distributions of the same material. Such bins may take the form of an open stockpile or closed vertical bins fed from the top. Front-end loaders or conveyor belts under a grating in the floor are the most common ways of bringing the open-stockpile materials into production. Mechanical feeders are used at the bottoms of closed bins to transfer the ground materials to conveyor belts leading to the forming machinery. A good bit of engineering design must go into the closed bins to insure positive, uninterrupted gravity feed. The feed from most closed bins is automatically and continuously weighed as it passes onto a conveyor belt. This insures the correct proportions of the various materials being blended into the desired composition [11].

Fig. 29. Additive hopper and feeder for introducing barium carbonate to the clay blend

Small additions of materials to the main clay blend are referred to as additives. Dry additives are fed from a hopper onto the conveyor belt taking the final blend to the mixing equipment, which is usually a pug mill. Each hopper is equipped with a calibrated and synchronized feeding device to give exactly the right amount of the additive. Fig. 29 is a photograph of such a feeder. The white chemical is added to the material on the conveyor belt by means of a disk feeder at the bottom of a hopper. There are other types of feeders also available and in use in this industry.

Some of the dry materials considered as additives to the clay batch are barium carbonate, soda ash, bentonite, colorants, and binders to increase the dry strength of the formed product.

Solutions or water suspension additives are fed from a liquid storage tank directly to the pug mill as part of the water added to obtain plasticity. Calcium chloride solutions, barium carbonate suspensions, and water-soluble binders are examples of materials added in liquid form. The proportions must be carefully controlled by means of automated valves and constant concentrations.

References

1. Dawson, A. S.: The application of geology in the search for industrial mineral deposits. Trans. Can. Inst. Mining and Met. 52, 23—26 (1949).
2. Ries, H., and T. L. Watson: Elements of Engineering Geology. New York: J. Wiley and Sons. 1947.
3. Pryor, E. J.: Mineral Processing, 3rd ed. New York: Elsevier Publishing Co. 1965.
4. Taggart, A. F.: Handbook of Mineral Dressing. New York: J. Wiley and Sons. 1945.
5. Testing Sieves, Handbook 53. Mentor, Ohio: The W. S. Tyler Co. 1969.
6. Lamar, R. S.: A review of methods for determining particle size distribution of ceramic raw materials. Am. Ceram. Soc. Bull. 31, 283—88 (1953).
7. Orr, C., Jr., and J. M. Dalla Valle: Fine Particle Measurement. New York: Macmillan Co. 1959.
8. Krumbein, W. C.: Size frequency distribution of sediments and the normal phi curve. J. Sed. Pet. 8, 84—90 (1938).
9. Brotherhood, G. R., and J. C. Griffiths: Mathematical derivation of the unique frequency curve. J. Sed. Pet. 17, 77—82 (1947).
10. Krumbein, W. C.: Size frequency distribution of sediments. J. Sed. Pet. 4, 65—77 (1934).
11. Johanson, J. R.: Method of calculating rate of discharge from hoppers and bins. Trans. Am. Inst. Mining, Met., Pet. Eng., 232SME, March, 1965; U.S. Steel Corp., Public Relations Dept., Tech. Paper 12 (26) 1965.

4. Forming of Structural Clay Products

4.1. Structure and Properties of Water

After preparation of the raw materials with regard to composition and particle size, structural clay products are usually formed in the plastic state. This means that water is added to the raw materials to produce the proper consistency and wet strength. In this process peculiar things happen that create the really unique *plasticity* of clays. We find that as water is added to dry clay there is as much or more of a change in the properties of the water as there appears to be in the alteration of the clay into a plastic, formable mass. For this reason it is necessary for us to pause here to look at the structure and properties of water before discussing the interactions between clay and water in an attempt to explain plasticity.

Seventy-one percent of the earth's surface is covered with water—in some places to a depth of several miles. We do not consider it an unusual substance. Yet on a scale as wide as the universe, water is relatively rare; indeed, the liquid state of matter, as we experience it, is not at all common. It appears that most of the material of the universe is either solid or gaseous, because each liquid exists only over a comparatively narrow temperature range. The naturally controlled temperature range existing on the surface of the earth makes it possible for us to observe as much of the liquid state of matter as we do. Water as a substance has many unique properties, but considered as a liquid, it is not particularly unusual. All liquids are different, one from another, in the same ways that solids are different.

Considering all the states of matter and their respective temperature-pressure relationships, one gets the impression that liquids are transition states between crystalline solids on one extreme and gases on the other. A closer study of the liquid state in the neighborhoods of the freezing points and the boiling points reveals that the liquid state of matter, and water in particular, is truly a transition state. Liquids, close to the freezing point, exhibit a great deal of the regularity characteristic of the parent crystalline solid, but at the boiling point, liquids have been described as being similar to a compressed gas [1].

In view of the fact that liquids are derived from crystalline solids by heating, we would expect to find binding forces and atomic arrangements in liquids similar to their individual crystalline forms. This resolves, then, into a classification of liquids with much the same parameters as are used to classify solids. There are ionic liquids such as molten salts, metallic liquids consisting of metallic atoms held together by extremely mobile electrons, molecular liquids in which the bonding is due to Van der Waal's forces, and liquids with a special type of bonding known as the hydrogen bond. Water is an example of the last.

Liquids, in general, have been classified as to *associated* and *nonassociated*. Forslind [2] has attempted to clarify this classification by stating that a nonassociated liquid is one in which the interaction between molecules is governed by fields having intensities with a constant angular distribution, while an associated liquid is one in which the distribution of angular intensities of the molecular fields is variable and directional. The interaction of molecules in a nonassociated liquid usually leads to a close-packed arrangement, but since the interaction of molecules takes place in preferred directions in the associated liquids, these liquids usually deviate from the close-packed state. Water is an example of an associated liquid with molecules having preferred bonding directions which give to the molecules a distinctive shape.

If the shape of the water molecule determines its chemical bonding and, in turn, the properties of water, then the structure of this molecule must be examined. Here another unique feature appears, as we find the electronic structure of the oxygen atom dominating the electronic structure of the whole water molecule. This situation is very unusual when one considers that the water molecule is made up of three atoms. Most complex molecules have electronic structures composed of molecular orbitals which show no resemblance to the electronic structure of any atom present. No doubt the unusual situation with water is due to the combination of an oxygen atom with two hydrogen atoms where each hydrogen atom contributes only one electron and one proton to the structure.

Let us look, then, at the electronic structure of the oxygen atom. Linnett and Poe [3] calculated the configuration of the electrons about the oxygen atom by statistical quantum mechanical procedures. The electrons involved with the oxygen nucleus are $1s^2$, $2s^2$, $2p_z^2$, $2p_x^1$, and $2p_y^1$. They assumed that the two K shell $(1s^2)$

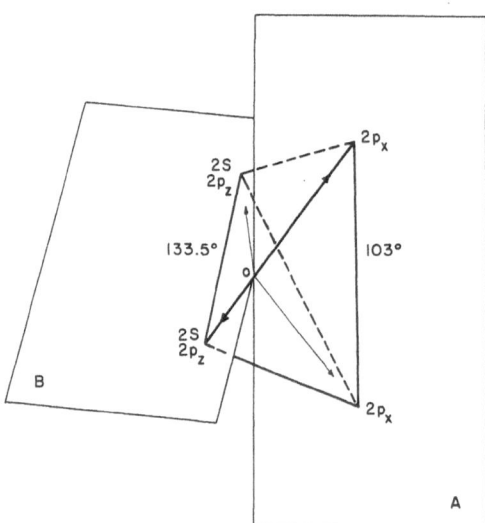

Fig. 30. Electronic configuration of the oxygen atom

electrons were symmetrically located about the oxygen $+8$ nucleus; therefore, they would not contribute to the characteristic shape of the atom. More precisely, the probability of finding the $1s$ electrons in the surface of a sphere around the nucleus was the greatest of all possible considerations. Since the L shell electrons, $2s$ and $2p$, have relatively small energy differences, they were found to be mixed in hybrid states. Two pairs of electrons and two individual electrons were calculated to be found most likely in four definite directions in space outward from the nucleus. As shown in Fig. 30, these four directional locations lie in two perpendicular planes, such as A and B. The intersection of these two planes at Point 0 represents the center of the oxygen nucleus. The angle between the two hybrid pairs of electrons lying in the horizontal plane B was determined to be $133.5°$, and the angle between the two individual electrons in the vertical plane A was found to be $103°$.

To construct the electronic configuration of the water molecule, all that is necessary is to place the two electrons from the two hydrogen atoms at the locations of the single $2p_x$ electrons where they become $2p_y$ electrons and add protons at the sites of these two pairs of electrons in the vertical plane, A. When this is done, the only change in the electronic configuration is a slight increase in the $103°$ angle to $104.7°$, caused by the repulsion force between the two positively charged protons. Once the molecule of water is formed, it presents nearly a tetrahedral (4-sided) shape in space, with a pair of electrons at two corners and protons at the other two corners. This provides highly negative influences at two corners and positive influences at the other two [4].

If we now put two water molecules together in a chemical bond, a positive point of one tetrahedron will be attracted to a negative corner of the other. When this contact is made, a hydrogen bond is established between the two tetrahedra, since the proton is now associated with two pairs of electrons. If several more water molecules are brought into the picture, each making hydrogen bonds at the points of contact, a hexagonal array is formed as is shown in the 2-dimensional drawing of Fig. 31.

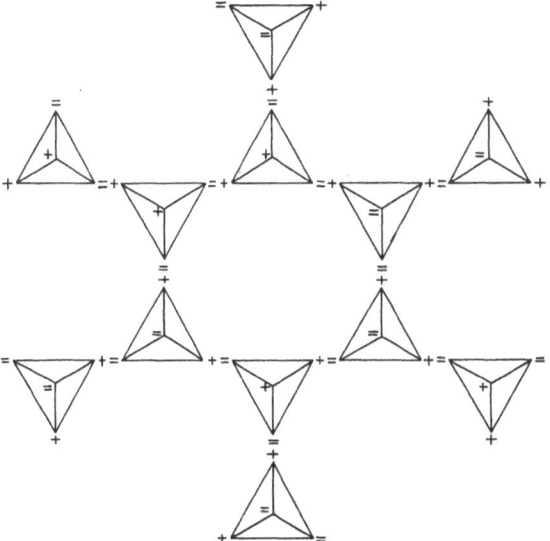

Fig. 31. Schematic ice structure

This drawing represents the crystal structure of ice, and one must realize that some of the unattached corners of the tetrahedra are pointed upward and others downward. Through these points the crystal is built into a 3-dimensional, hexagonal ice crystal. (Note the similarity of the ice structure to that of silicon-oxygen where the silica tetrahedra share corners as shown in Fig. 11.) The oxygen nuclei are located in the center of the tetrahedra representing water molecules. One can see from Fig. 31, the reason why snowflakes show beautiful hexagonal patterns.

It is more difficult to truly represent the structure of liquid water, since it is a dynamic, constantly changing array of molecules in a collapsed ice structure with many voids or vacant sites. Remembering that liquids are transitional states between the structure of the solid and that of the gas, it is clear that the structure of water will be different at various temperatures between 0°C and 100°C. In fact, Morgan and Warren [5] calculated from X-ray diffraction patterns of water the number of nearest neighbors and their separation distances from any given molecule, and found both the number and distances to increase with temperature as shown in Table 8.

Table 8. *Coordination of Water Molecules in Ice and Water* [5]

	1st Coordination		2nd Coordination	
	No. of Molecules	Distance (Å)	No. of Molecules	Distance (Å)
Ice	4	2.76	8	4.51
Water at 15°C	4.4	2.90	~12	4.5
Water at 83°C	4.9	3.05	completely random	

With all of these difficulties in mind, Fig. 32 is presented as a possible 2-dimensional, instantaneous arrangement of water molecules at some temperature not far above the freezing point. The collapsed hexagonal ice structure results in smaller empty spaces

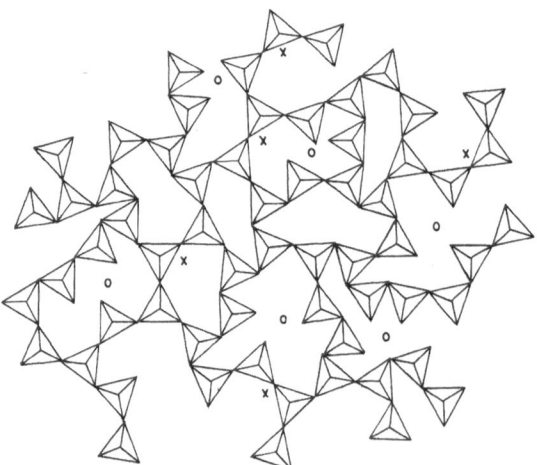

Fig. 32.   Schematic structure of water

until thermal expansion again places the molecules farther apart. This explains why ice floats on water and why water has its greatest density at 4°C. The X's on Fig. 32 call attention to a remnant of the original ice structure, and the O's locate the points where vacancies exist. (In making this drawing a schematic liberty was taken for the sake of clarity.)

Taking the structure of water as it has been described, what happens to the structure when a charged particle is introduced? For this situation, we might consider a solution of gaseous ions. One would expect, on first thought, that the freedom of movement of gaseous ions would be more restricted when placed in the liquid medium. Thermodynamically, this would be a loss of entropy (ΔS^0) of about 10 to 14 cal./deg./mole as shown in Table 9 under ΔS_i; however, something else of great importance happens because the total entropy losses are much greater than that. Total hydration entropies are listed under ΔS_h in Table 9 where entropy values are as high as -104 cal./deg./mole. The additional loss of entropy is caused by the ions structuring the water, so that many water molecules are less free to move about than they were before the ions were introduced. A glance at the ΔS_w column in Table 9 shows that the entropy loss due to the organization of water molecules out to some distance from each ion is greater than that for the restriction placed on gaseous ions in a water environment, except for K^+, Br^-, and I^-. Note that these are large ions with a single charge [6].

Table 9. *Entropy of Hydration of Gaseous Ions* (298°K) [6]

Ion	ΔS_h^0	ΔS_i	ΔS_w	Ionic Radius (Å)
Li^+	-34.7	-8.9	-25.8	0.68
Na^+	-28.3	-10.4	-17.9	0.98
K^+	-21.8	-11.3	-10.5	1.32
Mg^{2+}	-72.4	-10.5	-61.9	0.65
Ca^{2+}	-61.5	-11.4	-50.1	0.94
Ba^{2+}	-50.9	-14.4	-36.5	1.29
Al^{3+}	-104	-10.7	-93.3	0.47
Fe^{3+}	-98	-11.9	-86.1	0.56
F^-	-35.2	-10.0	-25.2	1.36
Cl^-	-24.2	-11.2	-13.0	1.81
Br^-	-15.4	-13.0	-2.4	1.95
I^-	-8.8	-14.1	$+5.3$	2.19

A closer look at the entropy values for the various ions in Table 9 reveals several interesting relations. First of all, the magnitude of the ionic charge has a profound effect on structuring the water. Li^+, Mg^{2+}, and Fe^{3+} have similar ionic radii, but their ΔS_w values are -25.8, -61.9, and -86.1, respectively. Secondly, the monovalent anions do not structure the water as much as the monovalent cations. In the third place, the effect of ionic size on the structuring of water is quite clear. Ions of the same charge have a decreasing structuring effect as their sizes increase. Notice that the iodide ion does not structure the water at all, but it seems to break up the normal structure of water into a more random arrangement. It appears, then, that *ions of small size and large charge have the greatest ability to impose a structure on the water molecules.*

The action of the ions to structure the water is caused by the attraction of the appropriate points of the tetrahedral molecules to the electrically charged ion; a long range structure is built up by dipole attraction of one water molecule after another which results in an "ice-like" structure surrounding each ion. If enough ions are present, the whole bulk of the water is affected.

If the theory of ionic structuring of the water is valid, properties such as viscosity and freezing point should demonstrate this effect because the tightening of the molecular structure should set up a resistance to flow and a difficulty for molecules to move into the crystalline ice structure. The solution of highly ionized salts is used to show these effects, and the data are presented in Table 10. Some anomalies can be found in this table probably due to the variations in the actual degree of ionization and the simultaneous introduction of both cations and anions.

Table 10. *Effects of Ionic Salts on Water Structure**

Salt (0.5 g-mol/1)	Freezing Point Depression ($^\circ$C)	Relative Viscosity (20°C)	Fluidity (20°C) (Poise^{-1}) rhe
KI	1.73	0.95	104
KBr	1.69	0.98	102
KCl	1.68	0.99	100
H_2O	0	1.0	99.8
NaCl	1.71	1.05	95
LiCl	1.81	1.08	93
$BaCl_2$	2.45	1.12	89
$SrCl_2$	2.56	1.14	87
$CaCl_2$	2.52	1.15	86
$MgCl_2$	2.71	1.22	81
$FeCl_3$	3.35	1.39	72

* Interpolated from Handbook of Chemistry and Physics, 55th Ed., CRC Press, Cleveland (1974—75)

There are many interesting effects brought out in Table 10 in spite of the anomalies. All salts listed depress the freezing point, but the alkali and alkaline earth chloride series again indicate that cations of smaller size have greater structuring effects. It has been noticed that concentrated solutions of $MgCl_2$, $AlCl_3$, and even NH_4OH when cooled by liquid nitrogen, form glasses rather than crystalline solids. One must assume, in these cases, that glass formation is the result of the decrease in mobility of the water molecules caused by the presence of foreign ions. The potassium halide series at the top of the table seems to indicate that anions have a smaller effect on freezing point depression than cations. The data on pure water was placed in this table to separate the salts which decrease viscosity and increase the fluidity of water, from those which give an opposite effect. You will recall that the entropy data from Table 9 showed that K^+, Br^-, and I^- ions structure the water slightly or not at all. A possible explanation for the decreases in fluidity with the potassium halides is that the ions have a greater tendency to break up the molecular network than to impose their own structure. More important for this discussion is the increase in viscosity and the decrease in fluidity for all the salts listed below H_2O. These data, together with the freezing point depressions, support the theory derived from entropy values that the introduction of charged particles causes a structuring effect on the water.

It will be shown presently that there is an analogy between the effects caused by the introduction of electrically charged ions into water and the introduction of tiny clay particles.

4.2. Clay-Water Interaction

When clay particles are placed in water the resulting properties can be explained only by assuming an interaction between the two. The clay grains are affected and altered in subtle ways, and the water structure is changed by the presence of the clay.

All inorganic crystalline solids have higher surface energies than is characteristic of their interiors. This is caused by the disruption of the repetitive crystal lattice at the surface, where cations and anions are left without their proper number of bonding neighbors. The discontinuity leaves surface ions seeking neighbors of some kind in their attempt to lower surface energy.

We will see many evidences that the unscreened surface cations contribute more to surface energy than the anions, because they are more demanding on their surroundings. This is because the electronic structure of the cations is less polarizable, owing to a deficiency of electrons around the positively charged nuclei, whereas the anions have excess electrons and can partially compensate for the lack of a neighbor by the distortion or polarization of their own electronic structures. Because of the cation effect, we find that the surfaces of crystals of quartz (SiO_2), corundum (Al_2O_3), and periclase (MgO) have high surface energies, but crystals of potassium iodide (KI), lead oxide (PbO), and cesium bromide (CsBr) have low surface energies. The high-surface-energy crystals have the greater power to adsorb foreign ions and molecules on their surfaces.

Relatively equidimensional crystals like those just mentioned can reduce their surface energies by withdrawing their exposed cations deeper into the interiors of the crystals. This leaves a substantial amorphous layer on the surface due to distortion of the crystal lattice by the cation displacement. We use this behavior in the determination of particle size by the X-ray diffraction method, and we observe it in the repulsive forces between very fine crystals of these compounds where the first surface layer of ions appears to be anions presenting an electrically negative influence. However, even with such surface distortion, there remains considerable surface energy from the increase in entropy; so the tendency to adsorb is still strong [7].

We can recall that in the crystal structure of clays we have highly charged, small cations (Si^{4+}, Al^{3+}, Mg^{2+}) in essentially a 2-dimensional lattice. In these crystals there is little room to self-create a distorted surface layer in an attempt to lower surface energy. It can be expected, then, that the surface of clay particles will be highly demanding on their environment as far as the power to adsorb foreign molecules is concerned. There will be unscreened strong cations which can act, in some respects, as ionic charges in solution when the clay particles are placed in water. There is a strong analogy between ions in solution and colloidal clay particles in suspension, which is evident in their effects on freezing point lowering, fluidity, pH, cation exchange, and the creation of an apparent electrical charge on the particles. When dealing with such small particles, the surface forces dominate the behavior of the whole because the amount of surface area relative to volume is very large.

When clay particles are placed in a water environment, ions are attracted to the surface sites; then, a structure of water molecules is built up around the particle. $(OH)^-$ ions are adsorbed onto the cationic sites and $(OH_3)^+$ ions on the anionic sites; although, they may be held at their respective sites by different bonding forces. The structured water hull is built up in a rigid, ice-like fashion under the influence of the surface sites. This is easily done by the water molecules orienting themselves so that hydrogen bonds exist throughout. The rigidity of the structured water gradually decreases, the farther the structure gets from the clay surface. There will be a point where water molecules are coming to and going from the hull. How far out this occurs depends on the thermal motion, which increases with temperature, and on such shear forces as might come from flowing or stirring the suspension.

It is interesting to note here that the hexagonal array of water molecules fits closely the hexagonal arrangement of the exposed silica and alumina sheets of the clay mineral layers. This good fit naturally promotes the water hull development [8]. Water hulls are built up around the crystallites of kaolinite and illite, but an additional effect is noticed with montmorillonite. In the case of montmorillonite, water forms a structure between the individual layers of the crystal, thereby causing an expansion of the crystal lattice in the c direction—perpendicular to the layers. Depending on the type of interlayer cation that may be present, the layers separate as more and more water is introduced into the system. For example, as the humidity of the air surrounding a clay particle of montmorillonite increases, the separation of the layers takes place, with pauses occurring for short intervals of humidity, increase when a water layer has been completed. As more water is added above 100% relative humidity, the layers continue to separate until, in some cases, the separation distance becomes too long to measure by conventional X-ray diffraction techniques.

Further evidence of clay particles structuring water molecules can be found in the depression of freezing points. When the concentration of clay in water is high enough, all of the water present is acted upon by the clay, and in this case the water exhibits either a depressed freezing point or no freezing point at all [9]. Fig. 33 shows the depression of the freezing point of water by three different clays. The percent of water on kaolinite where the water hulls are complete and free water is then present is clearly shown by the break in the curve at about 37%. The ultrafine particle size of montmorillonite and its expandable layers are responsible for the large amounts of water organized by bentonite [10]. In fact, a 10% deflocculated suspension of montmorillonite will structure the entire volume of water, and a type of gelation called thixotropy is witnessed.

Although clay particles obviously have no inherent charge in the dry state, they appear to develop electrical charges of varying magnitudes and sign when placed in water. The magnitudes of these charges on clay particles have been measured in electrolytic cells, and in this situation the colloidal clay particles behave like ions in

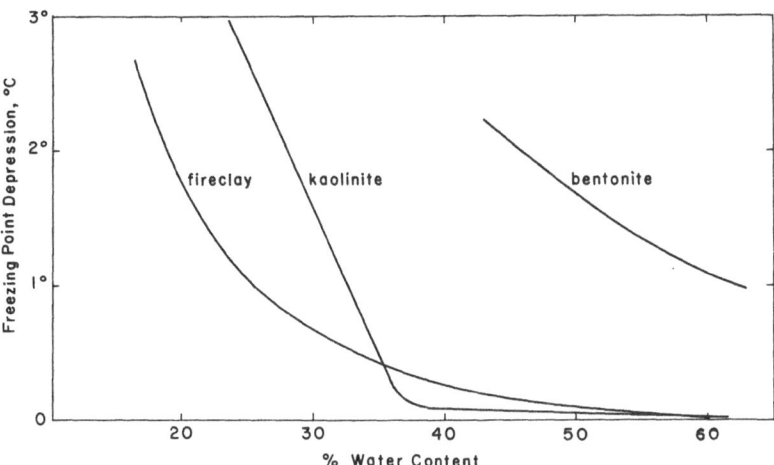

Fig. 33. Depression of the freezing point of water by clays. After Bodman and
        Day [10]

that they migrate under an electrical potential to the electrode of the opposite sign.
We are going to find that the charge developed on clay grains in water is related to the
differential adsorption of anions and cations.

To explain the origin and development of charges on clay particles in a water
medium, two theories have been proposed. The Gouy-Freundlich [11, 12] diffuse-
double-layer theory is used for the charges developed on colloids in general, and the
Lawrence [13] thermodynamic theory was developed especially for clay-water systems. Upon close examination, it will be found that both theories are based on the
same assumptions and both only approximate the realities of clay-water systems.

The theory of the diffuse electrical double layer as applied to clays assumes that
a fixed layer of $(OH)^-$ ions is attached to the surface of the particle, attracted there
by the unscreened cations; then a diffuse layer of counter ions $(OH_3)^+$ extending
some distance out from the particle is assumed. It is presumed that the concentration
of counter ions is greatest near the particle surface and gradually becomes lower as
the distance from the surface increases until the concentrations of both ions are
equal in the water medium. In the diffuse layer, it can be assumed that the concentration of $(OH)^-$ ions is less than in the normal water medium because of repulsive forces
from the adsorbed surface layer [14].

Clay particles in water, then, appear to have a negative charge because there are
more negative ions close to the particle than positive ones; however, in the whole
clay-water system the number of positive and negative ions is, of course, equal.
The clay particle and its water hull will move toward the anode if an electrical potential is applied as in an electrolytic cell. The imbalance of $(OH)^-$ ions between the
particle surface and some distance away where the concentrations of $(OH)^-$ and
$(OH_3)^+$ ions are equal, creates an electrical potential across this distance. The potential is called "zeta potential," but since it is defined by some unknown distance
away from the particle, it is not a useful quantitative measure of charge magnitude.
Qualitatively, the thickness of the diffuse layer is related to the zeta potential, which

also determines the charge on the particle. It is assumed that the apparent charge on the particle is large when the diffuse double layer is extensive, and the particle appears to have a low charge when the diffuse layer or zeta potential is small [14].

According to the double-layer theory, the negative charge on clay particles is small when the fixed layer is composed of $(OH)^-$ ions, and the diffuse layer contains a preponderance of $(OH_3)^+$ ions. This would also be true when highly charged cations, such as Ca^{2+}, Mg^{2+}, and Al^{3+}, are added to the system by way of an electrolyte, because the adsorbed negative charge is neutralized in a short distance by ions of larger charge. With a thin diffuse layer, particles can approach each other closely without repulsion; then Van der Waals forces of attraction exceed repulsive forces and particle aggregates are formed. This is called flocculation.

Deflocculation occurs in the diffuse layer theory when electrolytes are added which increase the thickness of the diffuse layer, the zeta potential, and the apparent charge on the particle. This is accomplished by adding monovalent ions such as Li^+ and Na^+ to the clay-water system. These ions are less able to neutralize the adsorbed $(OH)^-$ ions unless they are introduced in excessive quantities. In addition, they do not approach the surface as closely as $(OH_3)^+$ ions because of their lower mobilities. More precisely, they have lower mobilities than the proton. Deflocculation results from the charged particles repelling each other and approaching a stable colloid that does not settle out appreciably with time.

The Lawrence thermodynamic theory has some advantages over the double-layer concept. The thermodynamic approach provides a simple model for easy understanding and, of course, relates temperature to charge development. It is, unfortunately, completely independent of the size of the water hull or the structure built up around the clay particles.

In the Lawrence theory, the surfaces of the clay minerals are assumed to be composed of an array of partially coordinated cations and anions. The specific model was kaolinite where the partially exposed surface species were Si^{2+}, i.e., silicon with a residual charge of 2+, and $Al^{1.5+}$, and the anions were O^- and $(OH)^{0.5-}$. When the clay particle is placed in water, it is assumed that any cations in the water are attracted to the anion sites on the clay, and anions are attracted to the surface-cation sites. It is understood that the cations and anions attracted to the surface sites are adsorbed through different bonding energies. It is possible that thermal vibrations between the crystal site and the adsorbed species will occasionally break the bond, and at that instant, the site will be vacant. The following statistical equation is used to calculate the number of vacant sites of any kind at any temperature, and from these numbers, the charge on the particle can be calculated:

$$N = N_0 \exp\left(\frac{-E}{kT}\right) \tag{1}$$

where N is the number of vacant sites, N_0 the total number of sites available, E the bond energy between the site and the adsorbed ion, k the Boltzmann constant, and T the absolute temperature [13].

Lawrence calculated the bond energies of several cations and anions adsorbed on the typical sites of the kaolinite crystal lattice. In his calculation he assumed the cations in solution to be hydrated and the hydrated species to be adsorbed on the clay. The values he obtained and the corresponding values assuming the adsorbed ions were not hydrated are given in Table 11. The first important feature to note from these data are the relatively high bond energies of the (OH)⁻ ions on the crystal cationic sites and the low bond energies for the cations adsorbed on the oxygen sites.

Table 11. *Bond Energies of Ions on Clay Surfaces*

Ion	Bond Energy (ergs/bond$\times 10^{-12}$)	
	Hydrated Ion [13]	Unhydrated Ion
On oxygen sites:		
Li^+	4.6	10.8
Na^+	4.8	9.9
K^+	5.1	8.5
H_3O^+		8.5
NH_4^+	5.3	8.3
Mg^{+2}	8.0	21.6
Ca^{+2}	8.3	19.1
Al^{+3}	21.6	35.9
On Al sites:		
OH^-		18.0
On Si sites:		
OH^-		27.6

By inserting these values into Equation 1 and calculating the number of vacant sites, it becomes clear that clay particles should have negative charges. It is found that, in the temperature range of water, there are negligible vacant cationic sites. All of these sites are occupied by adsorbed (OH)⁻ ions. It is also found that a considerable number of surface anionic sites may be vacant, especially in the case of adsorbed Na^+ ions, where the bond energy is low. This differential adsorption of ions on the surface of clay results in a net negative charge due to the larger number of (OH)⁻ ions attached under specific thermal conditions. From the data of Table 11, it can now be appreciated that the magnitude and sign of the charge on clay particles vary with the types of cations adsorbed. The charge should be highly negative when monovalent cations are involved, slightly negative with divalent cations, and slightly positive with Al^{3+} under certain conditions. As a matter of fact, Button [14] observed the positive charge on kaolinite when saturated with aluminum ions.

The decision whether to use the hydrated cationic unit or the unhydrated cation in these calculations seems to be unclear with respect to observed behavior. Perhaps in some cases one value is more appropriate, and in other cases the other value proves to be better. Experience tells us that sodium clays have large negative charges and are deflocculated. If the values for unhydrated cations are used, potassium should provide greater negative charges, but this is not observed, which may be due to the unusually good geometric fit of potassium on the clay lattice providing a stronger bond than expected. The unhydrated divalent and trivalent cations give charge values more agreeable with observations.

Button and Lawrence [16] have shown that the charge on clay particles increases with temperature regardless of the type of cation adsorbed. As the temperature is increased, there is a tendency for the clay to go toward deflocculation with the higher charge. This is an important consideration because hot water is often used with clays in the structural clay products industry. It is important to know what this is doing to plasticity and the forming operation.

It is good to understand the mechanism of charge development on clay in order to be able to manipulate the properties resulting from it, and there are several practical aspects that come out of the theories just presented which are important to know. A clay-water system is deflocculated if the adsorbed ions are sodium; and calcium, hydrogen, and aluminum ions cause flocculation. Since there is a bonding preference for calcium over sodium ions, the exchange is essentially spontaneous when a salt such as $CaCl_2$ is added to a sodium-saturated clay. However, to exchange calcium ions adsorbed on clay for sodium ions requires a few tricks. In the laboratory either electrodialysis or the law of mass action can be employed, but in commercial production sodium salts that contain an anion which will render calcium ions insoluble must be used in small amounts. Salts of this kind are Na_2CO_3, Na_2SiO_3, and $(NaPO_3)_6$. The latter hexametaphosphate, is known under the trade name Calgon. In operation, the carbonate, silicate, or phosphate anions immediately precipitate calcium ions from the system as soon as these ions are kicked off the clay surface by thermal motion, and the sodium ion is readily present to take its place.

One usually thinks of flocculation and deflocculation of clays in terms of thin suspensions and slips, but they are of equal concern in the plastic state of clay-water mixtures. For clay suspensions of equal viscosity, the flocculated state requires more water than the deflocculated. In other words, flocculated clay-water systems tend to be more viscous, which means that if a suspension is made up to a particular consistency in the flocculated state, it will become more fluid if a deflocculating agent is added. The plastic state of clays responds in the same manner as slips with regard to flocculation and deflocculation. As a matter of fact, we are soon to learn that the difference in consistency of specific clay-water systems as one goes from the plastic state to slips is only due to the amount of free, unstructured, water present. In the plastic state the free water is only a thin film about each particle, while in thinner suspensions the free-water volume separating the grains may be much larger.

The charge on clay particles in water determines the state of aggregation which, in turn, affects the consistency and plasticity.

4.3. Plasticity of Clays

Since nearly all structural clay products are formed from clay materials in the plastic state and none are made by slip casting, we shall concern ourselves especially with the plasticity of clay-water systems. The plasticity of clays has been a difficult property to define precisely—perhaps because everyone has been seeking to apply a general mechanical definition to the very special behavior found in clay-water mixtures. In the past we have used the words workability, consistency, extrudability, and plasticity somewhat interchangeably, always with some reservations that they are not synonymous and that we may, at times, be using the wrong word to express ourselves adequately.

For reasons that will be clear shortly, these words will be defined for our particular use as they apply only to clay-water systems. This is necessary because clays have so many unique properties and behaviors it is impossible to encompass all in a single definition; however, if some of these features are understood when we are thinking of clay materials, the major properties can be described in rather simple terms.

Workability is a workman's term, and it refers to the ease with which a moist clay mass conforms to the particular process or job at hand. In each case the gradations may be: excellent, good, fair, or poor workability; although, the same clay paste may be graded differently from task to task and even, perhaps, from one workman to another. Good workability is probably related to the ability of the clay mixture to develop satisfactory forming properties over a relatively wide range of water contents.

Consistency of a clay-water system is simply its resistance to shearing forces such as in stirring. Of course it is greatly dependent on water content. Any clay has low consistency at very high and very low water contents. Presently we are going to see that consistency goes through a maximum as the water content is increased from zero to excessively high values. Consistency has little relation to the shaping or workability of a clay mass.

Extrudability is a term often used in the industry where extrusion is the method of forming, and it refers to the ease of extruding a clay mixture through a die. It is not closely related to consistency, because a clay material will extrude easily at maximum consistency, but if the consistency is lowered by the removal of water the extrudability may become very difficult. On the other hand, if the maximum consistency is lowered by the addition of water the ease of extrusion increases.

Plasticity is to be defined in more precise terms so that it will be more useful in scientific studies on the behavior of clay-water systems. *Plasticity is the ability of a clay-water mass at its maximum consistency to be shaped and to hold its shape after the forming forces are removed* [17]. The word ability in this definition has two connotations—one with respect to shaping and the other to holding the shape. In the first case ability refers to the amount of plastic or viscous flow (strain) that can take place before rupture occurs. In the second instance ability refers to the internal strength of the plastic mass which will be great when little deformation occurs after the forming stresses are removed. This dual relation in the definition qualitatively means that a clay will have high plasticity when a large force is required to deform it, and in this condition it will be deformed to a considerable degree before failure. As a matter of fact, a clay-water system of high plasticity requires more force to deform it and deforms to a greater extent without cracking than one of low plasticity which deforms more easily and ruptures sooner.

There are many factors affecting the plasticity of clay bodies, and it would be well to thoroughly understand these factors in order to be able to manipulate this property. Some of the factors are primary and others are secondary. The primary factors must all be present to achieve plasticity while the secondary factors can be used to control or regulate plasticity once it is present. The primary factors are:

1. the anisodimensional shape of clay-mineral particles,
2. the strong surface forces on clay-mineral particles due to incompletely coordinated small cations with large charge,
3. the rigidity of the water structure surrounding the clay mineral particles.
The secondary factors affecting plasticity do not create or destroy plasticity of clay bodies but only increase or decrease its magnitude. These secondary factors are:
a. the type of cations adsorbed on the clay-mineral particles and/or contained in their water hulls,
b. the temperature of the clay-water system,
c. the small particle size of the clay-mineral particles,
d. the presence of nonclay particles in the clay-containing body.

Clay bodies are truly unique, as was previously mentioned, in that they exhibit a group of properties and behaviors not all of which are present in other systems. These properties are:

1. Clay bodies develop plasticity as previously defined.
2. The plastic mass retains a large volume of water.
3. Clay bodies shrink on drying.
4. The strength of clay bodies increases on drying.

Other fine-grained materials do not develop this group of properties. For example, a finely ground quartz-water system has no plasticity. It has some consistency and can be shaped, but the piece will not hold its shape after the forming pressure is removed. A quartz-water system will not retain much water, since most of the water added will quickly run out of the formed piece. The mixture will not shrink on drying, and it loses all its strength and cohesiveness on drying.

A good way to observe the development of plasticity of a clay material as water is added to the system is with the Brabender plastograph. This instrument measures and records the consistency of the material by means of the torque developed on the electric drive motor connected to a mixing chamber that resembles a double-shaft pug mill. The two rotating shafts carry rods, knives, or blades that act to stir the sample. After introducing the dry powdered sample, the machine is balanced to read zero consistency; then water is added at a constant rate, usually from 0.5 to 1.0 ml/min.

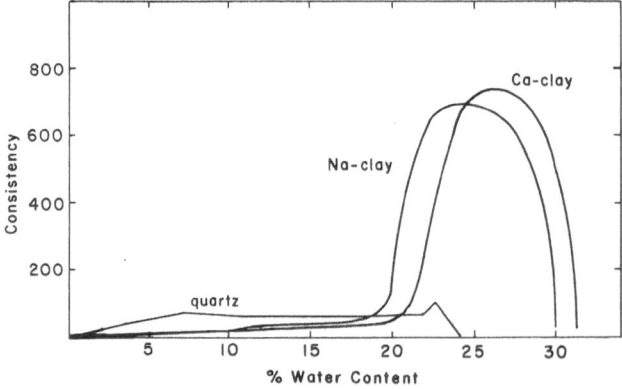

Fig. 34. Consistency development as water is added to clay and pulverized quartz

Fig. 34 shows three plastograph curves. Two curves are for clays, and one is for a nonplastic, pulverized quartz. Note that little or no increase in consistency occurs as water is added to the clays until the water content reaches a relatively high value. During this period water is being adsorbed on the surfaces of the clay grains. First the rigid layers of water molecules are built up about each grain; then, gradually, less-rigid water hulls are developed. The water thus adsorbed, is no longer free liquid water, and it is unable to flow. The clay-water system in this state appears dry until complete water layers surround each particle. No cohesiveness is developed between particles because, for all intents and purposes, the clay grains with their adsorbed water layers are solid particles [17, 18, 19].

Above about 18% water the clay curves of Fig. 34 rise rapidly to a maximum consistency within the short range of an additional 6% or 7% water. Before this period the water hulls have been completed, and they have come into equilibrium thickness with the shear rates imposed by the machine. Free water is now introduced between the grains, and the capillary attraction, due to the surface tension of the free water medium, is creating a cohesiveness between particles which increases up to the point of maximum consistency [20, 21]. As additional water is added past the maximum, the thickness of the free water between particles becomes larger and larger until the capillary attraction disappears, the clay-water system turns into a slip, and the consistency returns to zero as previously defined in the experiment.

Note the curve for the finely ground quartz. By our definition it has no plasticity, but a low-level consistency is developed in this material with just a little water. Probably the capillary attractive forces maintain the consistency over a wide range of water contents. Because of this great difference in behavior, one might consider the consistency measured on clays as being close to, but not quite, a measure of plasticity. At least in comparing one clay with another, one can say that the clay which develops the greatest consistency is the more plastic.

It should be kept in mind that the plastograph test is usually dynamic in nature and clay-water equilibrium distribution is not obtained. For this reason, the water content values at the points of maximum consistency tend to be higher for equilibrium conditions. There is little doubt that equilibrium mixtures are more nearly approached in the factory than in the plastograph test just described.

The effect of charge development on the consistency of clays is also shown on Fig. 34 where two plastograph curves of the same clay have been drawn. The maximum consistency of the sodium clay is lower and is reached at a lower water content than with the calcium clay. According to our definition of plasticity, one could say that the Ca-clay is more plastic than the Na-clay.

In the structural clay products industry, high plasticity or high internal strength is usually desirable; therefore, to optimize the plastic strength, the clay-mineral particles should be in a flocculated state or, in other words, have a very low charge. When more than enough plastic strength is available due to the nature and amount of clay present, some plasticity can be sacrificed by deflocculation for the advantages of less forming water and lower energy requirements for extrusion and drying. If the plastic strength is already low, this compromise may not be possible without severe distortion and cracking of the product. Plasticity, then, can be adjusted within narrow limits by the type of cations adsorbed on the clay grains.

Another method of modifying plasticity in practice is by temperature. You will recall that the number of cation vacancies, and, therefore, the charge on clay particles increases with temperature according to Equation (4.1). So, as the temperature of any particular clay-water system is increased, plasticity decreases. In addition to the charge development, an increase in temperature lowers the rigidity and size of the adsorbed water hull by increased thermal motion, and it decreases the surface tension of the water medium which tends to lower the capillary cohesion [22]. These effects of temperature on the consistency of clay have been demonstrated by West [23] using the Brabender plastograph which is equipped for the control of temperature of the clay being tested. In his work the maximum consistency was over 1000 at $0°$, 950 at $30°$, and 720 at $70°C$. This is a 28% loss in plastic strength due to a temperature increase alone.

In the structural clay products industry some forming operations, especially in extrusion, use hot water for tempering the clay material. This can be done when more plastic strength is derived from the mineralogical composition of the raw material than is actually required. Hot water should not be used where the inherent plasticity of the clay raw materials is already low. The advantages of using hot water in the extrusion process are exactly what would be expected from lowering the plasticity. They are: lower extrusion pressure, lower power required, lower water content, faster drying, and less energy consumed for drying. The disadvantage is that distortion and cracking can occur due to the loss of plastic strength.

The ability of a plastic clay body to retain its shape after the forming forces are removed has not been adequately or completely explained as yet. You will recall that this was part of the definition of plasticity. The published literature on the plasticity of clays is tremendous and confusing. Empirical behavioral data have been reported in every conceivable way, but the experiments were often performed without proper controls; therefore, inconsistencies in data are present, and the conclusions drawn may be of no value. Plasticity has been studied by compression, tension, torque, and extrusion, and different behaviors seem to be obtained from one kind of stress application to another. This has been partly due to greater sensitivity from one apparatus to another both in design of the test machinery and in the application of the type of stress most appropriate for plastic flow in clay. Inconsistent starting points for studies on plasticity have also been the source of some of the variations.

In spite of the confusion on clay plasticity in the literature, a careful scrutiny of past experiences brings forth several concepts which seem to be most agreeable. As a matter of fact the following three principles may be considered as well established:

1. As a plastic clay is deformed, it requires increasing force to continue deformation up to the point of failure.

2. As the rate of application of stress increases, the amount of deformation at any particular load decreases. The plastic mass becomes stronger the more rapidly it is deformed, and this even applies to the ultimate strength before rupture. Note that this appears to be an increase in plasticity with an increase in the rate of applied force, if we consider our original definition of plasticity.

3. Plastic flow of clay-water systems is not Newtonian, that is, stress is not directly proportional to strain.

Macey [24] was one of the first to report on two significant behavioral patterns of clay pastes which are relevant to understanding plasticity. (1) After observing that the slightest stress caused flow in a plastic clay-water mixture, he contended that no consistent and definable yield point exists except if one adopts the definition that the yield point is that minimum stress at which continuous flow takes place. This type of yield point has little practical value. (2) To explain the fact that when a constant load is applied to a plastic clay specimen deformation will occur to a certain extent, then cease until a greater load is applied (Principle No. 1), he brought forth the concept of strain hardening. He explained shear hardening as the movement of clay particles into a more stable arrangement, and the development of slip planes in the piece. From this, Macey recognized that the preparation and handling of specimens before testing was extremely important for reproducibility of experimental results. The uniformity that he advocated as necessary was not realized in his work.

The observations by Macey were elaborated by Bloor [25] when he recognized that strain hardening of clay pastes was caused by the alignment of the plate-like clay grains by plastic flow in such a way as to resist the flow or deformation. He theorized that the flat flakes moved during deformation so that their flat surfaces became perpendicular to the direction of applied force. Bloor [26] later demonstrated that clay grains do, in fact, become aligned in this manner during plastic deformation.

Particle orientation to resist applied forces explains a number of the properties observed in plastic clay masses. The fact that yield points are difficult to define, and yet something resembling a yield point is often observed in clay testing, is caused by the prealignment of the clay particles in the paste by the process used to prepare specimens for testing; so, they already have a history of deformation before the testing begins.

The alignment of clay particles adds to the explanation of why plastic clay bodies hold their shapes after the forming pressures are released, since the slightest deformation, even by the force of gravity, tends to strengthen the pieces and to cause them to resist further deformation. This action is, of course, in addition to the cohesive forces of the capillary water in the plastic clay mass.

Buessem and Nagy [27] were the first to observe the plastic behavior of clay starting with specimens having completely random orientation of clay particles. By a special method of forming, their cylindrical specimens had a random arrangement of clay particles as verified by X-ray diffraction. They used compression forces to obtain stress-strain curves at a constant rate of deformation on plastic kaolin specimens. During the tests, lubricated glass plates were placed against the ends of the cylinders to achieve uniform compressive forces. The significant behaviors which they observed are sketched on Figs. 35 and 36. An important feature of these curves is that they all start at zero; therefore, there was no elastic yield point. The slightest stress produced some deformation.

The property of plastic clays to demonstrate greater plasticity at higher rates of deformation is well established by the curves of Fig. 35. This was mentioned earlier as Principle No. 2 because many investigators over the years have observed this behavior. Apparently, the faster rates do not allow time for stressed particles to move in such a way as to find the easiest path to their equilibrium positions.

The necessity for starting plasticity measurement tests with randomly oriented specimens is brought out in Fig. 36. Buessem and Nagy obtained curves like these by

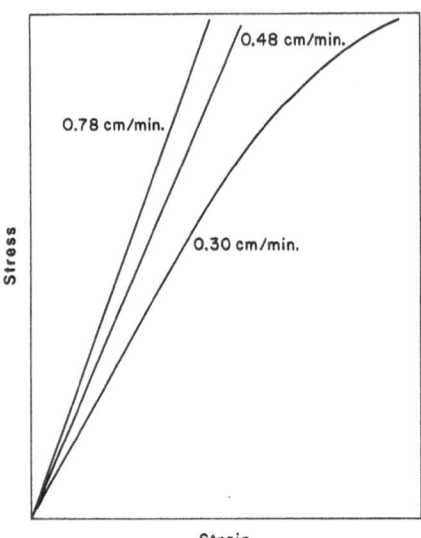

Fig. 35. Effect of rate of deformation on the plasticity of clay. After Buessem and
Nagy [27]

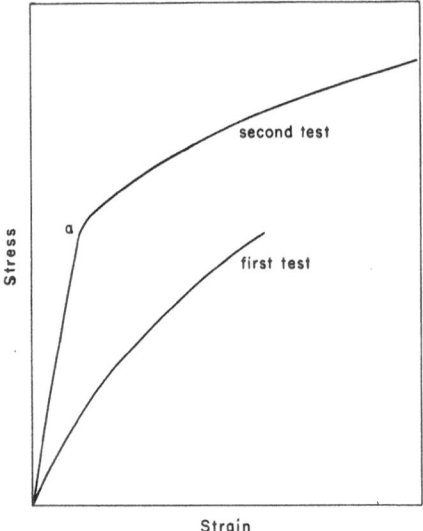

Fig. 36. Effect of predeformation on the plastic behavior of clay. After Buessem and
Nagy [27]

deforming a randomly oriented specimen a small amount, relieving the load, and
starting load application again from the zero point. It can be seen that the piece is
stronger after predeformation causing a resistive orientation of the clay grains up
to the load which had been previously applied. After passing the point of the maximum first deforming load, the deformation occurs at the same rate as would be
expected if the initial test had been carried out to greater deformation. This is a classic

example of strain hardening, and the break at Point (a) is an example of the yield points that many investigators have seen and tried to explain. It is obvious now that the value of this yield point is strictly dependent on the deformation history of the specimen tested.

More recent compressive tests on a plastic clay body were made by Kellogg and Sonneville [28] in a manner similar to that of Buessem and Nagy. The exceptions to the procedure were that Kellogg and Sonneville conducted their tests on extruded cylinders that contained considerable clay-particle alignment, and they did not use lubricated plates to insure continuous uniform loading. These differences in procedures are clearly visible on the curves of Fig. 37. The yield points are typical of those

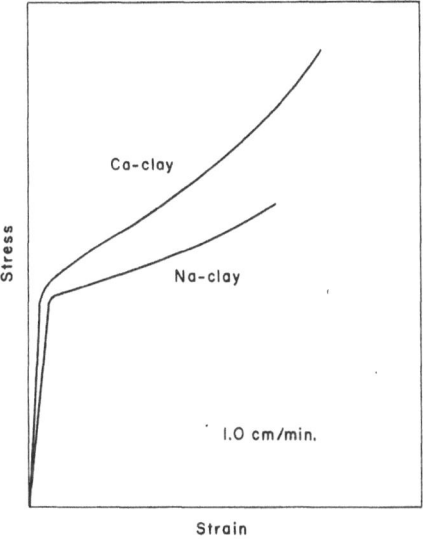

Fig. 37. Effect of adsorbed cations on the plastic behavior of clay. After Kellogg and Sonneville [28]

caused by prestressing the test specimens during preparation to cause particle orientation. The other interesting feature to compare with Buessem and Nagy's work is the continuous increase in slope of the curves above the yield point. This may be due to a reverse orientation of slip planes, the nonuniform distribution of the load during the test, or the presence of a nonplastic fraction in the Kellogg-Sonneville clay.

These inconsistencies do not alter the real reason for including Fig. 37, which is to show the effect of cations on plastic behavior. As might be expected from having seen the plastograph curves of these clays in Fig. 34, the Ca-clay shows greater plasticity and greater deformation before rupture than the Na-clay. Both curves were terminated at the first visible crack or failure of the specimen. The ionic effects can be explained by the lower charge development and the consequent flocculation of the Ca-clay and the fact that calcium ions have a greater structuring effect on the water hulls and the capillary water holding the particles together.

Before concluding this discussion on the theory of plasticity of clays, reference should be made to the work of Astbury [29]. He has produced a mathematical model of plastic clay being deformed by a sinusoidal application of torque. This model has not produced practical results that can readily be used in the structural clay products industry, but for those who wish to follow up on the study of the plastic behavior of clays, his work should be studied carefully and expanded into the realm of practicality.

4.4. Plastic Forming Methods

4.4.1. Plasticity in Forming

One of the most pressing needs in the forming of structural clay products is raw materials that develop high plasticity, because the ware must often withstand great compressive and shear loads in the plastic state. Extruded bricks and structural tiles are often stacked on kiln cars many courses high while in the plastic state as shown in Fig. 38. Such stacking places compressive and shear forces on the bottom courses, and if the ware does not possess sufficient plastic strength, cracking will occur. Cracks

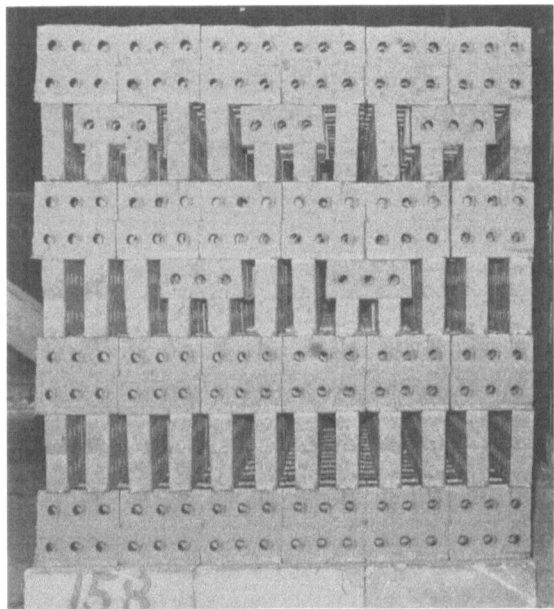

Fig. 38. Stack of face bricks on a kiln car

develop sometimes only in the lowest courses, but in other cases, they may extend upward into the ninth or tenth layer. In any case these scrap losses are costly and should be avoided for a successful business. Clay sewer pipes are not stacked like bricks, but some of the pipes are very large and heavy as can be seen in Fig. 39.

Fig. 39. Large bell-and-spigot sewer pipe being set upright for drying

They exert their own compressive forces that tend to distort and crack the lower ends if sufficient plasticity is lacking.

4.4.2. Soft-Mud Process

In the soft-mud brick manufacturing process, plasticity is not as much of a factor as workability and consistency. Soft wet clay must be packed into damp wooden molds that have been previously dusted with sand. It is necessary for the clay to flow easily into the shape of the mold with low-pressure packing. In the soft-mud process, the bricks are not stacked in the wet condition, but the molds are emptied onto

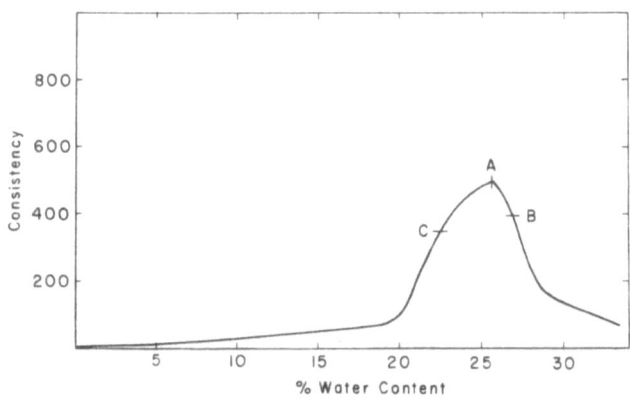

Fig. 40. Plastograph of a pulverized shale

pallets where the bricks stay until dried. The properties of workability and consistency are achieved by adding water beyond the amount required for maximum consistency; thus, the term "soft mud" is quite appropriate. The range of water contents for soft-mud brick production is nicely laid out on the plastograph of Fig. 40 which was made on a pulverized shale. (Some differences will be noted between the shape of this curve and those of the clay in Fig. 34.) The consistency range for soft-mud brick production is from Point A to Point B—on the side of the shale curve going towards a slip condition.

As mentioned earlier, the soft-mud process is an automation of the old technique of hand-molding bricks. An automatic brick machine of this type is shown in Fig. 41.

Fig. 41. Soft-mud brick machine

The mixing and tempering tub or pug mill is visible on the right-hand side just forward of the main drive mechanism. The soft plastic clay is pushed from the pug mill into the main part of the machine where it is proportioned and pressed into the mold boxes. The mold boxes are tied together in a gang of from 8 to 12 units. As the mold boxes emerge from the machine at the left, the excess clay is scraped off the top. At the extreme left in the photograph, the gang molds are tipped over and tapped to release the bricks onto a waiting pallet. As the loaded pallet goes to a dryer rack, the empty molds move below the machine to the right where they are washed with a water spray and resanded. They are then elevated into the machine ready to be filled again.

The sand used for lining the molds serves two purposes. It allows for easy release of the sticky clay from the molds, and it is a decorative feature. The basic sand is usually fine grained (perhaps −40 mesh), but sometimes sands as coarse as −10 mesh are used. The fired color of the sand and the texture it imposes on the bricks determine the appearance of the finished products. The mold sands work best in this machinery if they are thoroughly dry; therefore, a soft-mud brick plant has a sand preparation department where the sands are dried, screened, and mixed in preparation for delivery to the brickmaking machine.

Soft-mud facing bricks have a distinctive appearance. They are coated with sand on five sides, and the sixth side shows where the clay was scraped off. This scraping action always leaves a burr along the edge of the bricks as can be seen in Fig. 42.

Fig. 42. A typical soft-mud brick

These burrs are along the face of the bricks; so they can be observed even when placed in a wall. You can see that the corners and edges are not true, and the bottom corners are usually rounded where the clay did not fill the mold cavity completely. In dumping the bricks out of the molds onto the pallets, the unsanded side is down, and this action sometimes causes the soft clay piece to slump slightly. This makes the area of the unsanded side larger than its opposite sanded side. The shape of the whole brick is, then, somewhat truncated. These deviations from true shape are esthetically desirable for soft-mud products. The bricks not only appear to be antique, but the lack of perfection produces an interesting wall which might otherwise be monotonous. There is nothing about the forming of these bricks which detracts from their serviceability. Durability is related to the raw materials and the firing operation.

4.4.3. Stiff-Mud Process

The shapes of structural clay products made by the stiff-mud process are precise and uniform, and the standards for the products made by this process require them to be so. For these reasons, stiff-mud forming demands the greatest application of the fundamentals of clay-water systems discussed earlier, to optimize the process with regard to product tolerances and economics. The plastic state or plasticity of clay-water systems is the key to understanding the process well enough to produce the best results. One must be able to regulate the degree of plasticity through the secondary factors while recognizing the requirements for the primary factors.

Most structural clay products, including bricks, structural tiles, square and rectangular quarry tiles, and pipes, are made by the extrusion process. Machinery for extrusion is adaptable to automation and high speed production. Extrusion machines are available in several sizes. Some are capable of producing products at the rate of 150,000 standard brick equivalents per day, and they require electric motors ranging from 75 H.P. to 350 H.P. Typical extrusion machines as supplied to the structural clay products industry are shown in Figs. 43, 44, and 45.

Fig. 43. Stiff-mud extrusion machine showing auger which forces clay through a die. Courtesy of Fate-Root-Heath Company

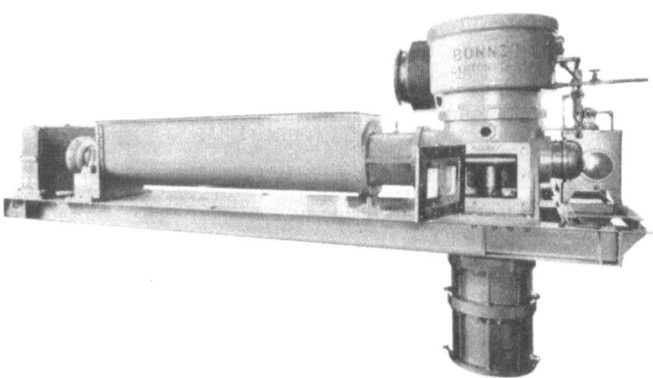

Fig. 44. Vertical auger extrusion machine for forming pipes. Courtesy of The Bonnot Company

All three of the machines just mentioned show pug mill tubs to the left in which the clay materials and additives are blended and the mixture tempered with water to the desired consistency. A view into a pug mill during operation is given in Fig. 46 where the mixing knives can be seen. Different types of knives are available to suit various raw materials, and the knives have variable pitches for the purpose of varying the time that the clay remains in the mill. The additives that may affect plasticity are

Fig. 45. Stiff-mud extrusion machine showing arrangement for quick-change dies. Courtesy of J. C. Steele and Sons

Fig. 46. Pug mill blending and tempering a clay mix

binders, internal lubricants, and wetting agents. Lignosulfonate binders are commonly used to increase the strength of the product after drying, but they usually act to lower the plastic strength by acting as a deflocculant. Water-soluble or emulsified internal lubricants allow for easier extrusion but reduce plasticity by interfering with the structured water hulls. Wetting agents act to reduce the surface tension of the water, which allows less water to be added, reduces the power necessary for extrusion, and speeds up the drying time, but all of these advantages are gained at the expense of plasticity.

Wetting agents reduce capillary attractive forces and often act as deflocculants. In most factories, the use of a lignosulfonate binder as a pug-mill additive is found to be beneficial overall in spite of its effect on plasticity [30, 31, 32].

Although theoretically products should be extruded at the optimum water content for maximum plastic strength, some plants extrude a somewhat drier mixture. This places the practical range of water contents for extrusion between Points A and C on Fig. 40 for a typical raw material. This is on the dry side of the maximum on the plastograph curve. Mixtures with water contents lower than the optimum can be extruded only when more plastic strength is available than is necessary to prevent cracking.

It may be economically desirable to extrude as dry as possible and sacrifice some plastic strength. The economics are questionable, however, when one observes the extra power required to extrude a drier column. The advantages to a fairly dry extrusion were reported by Hodgkinson [33] and they are:

1. The harder product is more amenable to automatic handling. (It must be remembered that this harder product cannot be deformed in the slightest without cracking, because of its lack of plasticity.)

2. The lower water content saves fuel on drying.

3. Better quality and shape of product is obtained.

4. The ware can be stacked on cars for drying so that pallet drying is not required. (Ware can only be stacked when enough plastic strength is still available to prevent cracking under heavy load. It is questionable, with the present state of automation, whether stacking wet bricks on a car is an advantage over pallet drying procedures.)

At whatever water content works best in a particular factory, the steepness of the plastograph curves indicates the necessity for close moisture control. Two or three percent water variation makes a large difference in consistency. Some materials have broader plastograph peaks than others, and these materials give slightly greater latitude in moisture content.

Up to about 1960, the control of water added was manual. Some operators watched a wattmeter or an ammeter connected to the motor driving the extrusion machine as a guide to consistency. Others used only their feel of the stiffness of the extruded column as a guide. Such manual control of water content has been monitored, and it was found that differences in water contents were negligible from one 10 or 15 minute period to the next, but a large sinusoidal variation with a wave length of 2 to 3 hours went through a day's production. For example, shortly after starting up in the morning, the water content was correct. As the morning wore on, the clay column became wetter until just before lunch a soft column would be recognized and the proper correction applied. Toward midafternoon the column became obviously too dry, and steps were taken to get more water into the mill. At the day's end, the extruded product was headed toward a too-wet condition. This total variation was 4%; 2% above and 2% below the desired water content.

In the 1960's, reports were being published on various devices to automatically control moisture content in the pugging process, and all produced smaller variations than manual control. Tatnall [34] placed a load cell on the tie rod of the auger section of the extrusion machine to transmit a hydraulic signal, based on the force required to extrude the clay, to a proportional controller that operated a pneumatic control valve on the water supply. In order to make this system operate properly, the raw material fed to the pug mill was volumetrically controlled, with a pneumatic gate valve regulating the flow to a constant-speed conveyor belt. It was also necessary to have a constant-pressure water supply. A control system based on the electrical resistance across the pugged clay in the mill was perfected by Connor [35]. (Electrical resistance increases with a decrease in water content.) The controls were actuated by the amount of electrical imbalance around a control point detected by the instrument. Proportional control was used to regulate up to 20% of the total water input. The bulk of the water was fed continuously into the mill at a constant rate.

The Consistodyne system developed jointly by Leeds & Northrup Company and the National Clay Pipe Research Corporation is now widely used in the industry to control water content in extrusion machines. This system employes two force detectors to "feel" the clay consistency by direct contact with the pugged clay. Probes with strain gages at their bases are installed through the sealing-die case directly into the flow of plastic clay at the downstream end of the auger. The detectors react to the stiffness by transmitting signals by way of a proportional controller to a mechanized valve.

After the mixed and tempered clay leaves the pug mill, the clay is forced through a sealing ring into a vacuum-auger chamber as shown just to the right of the pug in Figs. 43, 44, and 45. A high-capacity vacuum pump must be hooked up to this part of the extrusion machine. The sealing ring acts to close the upper vacuum chamber, and the die full of clay becomes the lower vacuum seal. As the plastic clay enters the vacuum chamber, it encounters a shredder which breaks up the compacted column to facilitate the deairing process. The lumps of clay from the shredder fall on the lower main extrusion auger. This auger is visible at the right end of the machine shown in Fig. 43. The auger is mounted vertically in Fig. 44, for extruding the clay downward.

The auger screw is the driving force to push the deaired clay through the forming die. The helix angle has been found to be best at 20° to 25°. Some augers are single-wing and others are double. The double-wing screw moves the clay more rapidly per revolution and keeps a more uniform pressure on the die. At the downward end of the auger screw, a propeller is sometimes mounted to break up the continuity of the smooth cuts in the clay mass coming off the auger flights.

One of the problems with auger extrusion is that the die puts a great deal of back pressure on the auger; so some of the clay tends to be forced backward off the ends of the auger screw. These extrusion pressures vary from 500 psi to 1500 psi (35 to 105.5 kg/mm^2), and the actual operating pressure is determined by the auger feed and the cross-section of the extruded shape formed by the die. Both of these extremes in pressure are undesirable, and probably the optimum extrusion pressure is around 800 psi (56 kg/cm^2), achieved with a proper auger-feed: die-size ratio. A good arrangement is to have the volume of clay extruded equal to or slightly greater than 50% of the total auger capacity. This indicates that to minimize backflow, the auger should not be run at full capacity. It has been recommended by the equipment manufacturers that the auger be operated at 40% to 60% of capacity. In the industry this is called a slightly starved auger. Backflow is also retarded by making the inside of the auger barrel rough, even to the extent of adding ridges to increase the resistance to flow [36].

In these machines there is an optimum extruded-column speed which is also related to the auger feed and die cross-section, as far as auger pressure and backflow is concerned. Column speeds of around 20 ft/min. (6 m/min.) are considered too slow, and 70 ft/min. (21 m/min.) is too fast. Obviously, this places the optimum column speed around 40 ft/min. (12 m/min.) for the best balance of all these factors.

As the plastic clay leaves the auger it enters the throat, or spacer, which leads the plastic clay into the die. The spacer can be seen just behind the mounted die of Fig. 45. The proper length and taper of the spacer seem to depend on the plastic characteristics of the clay material, to obtain the best product at the lowest power requirement; however, quantitative specifications for this part of the machine have not been agreed upon. One thing must be kept in mind is that a single-wing auger would require a throat section about twice as long as a double-wing [36].

A brick die can be seen mounted on the front of the extrusion machine in Fig. 45. You will notice that the mounted die is hinged on the left-hand side and another die can be mounted on the collar hinged to the right. This is a quick-change die arrangement. During production a different-shaped die can be installed in a matter of minutes by unbolting the die and swinging it aside, allowing the new die to be swung into place and bolted. The unused die can then be removed from the collar for cleaning and storage. The next die to be used can be mounted on the collar while the machine is in production.

The length of the die for all extruded products should not be any longer than necessary for obtaining a well-formed column. A rule of thumb for die length is 1.5 to 2 times the greatest distance across the die or 3 to 4 times the wall thickness of the piece being extruded [36].

One of the jobs of the plant engineer is to make certain that the die is centered on the auger and balanced. The flow of plastic clay must be uniform through the die side-to-side, top-to-bottom, and outside-to-inside. The flow through the die can be checked by cutting off the extruded column even with the outer edge of the die and running out a short column for observation. A die off center to the right will extrude a circular column to the right. A more detailed analysis of the column flow can be made by placing vertical and horizontal cutting wires in front of the die so that the extruded test column will be cut into four parts. The four separate columns will give more information on the places where the column is moving faster or slower. When alignment is good, further adjustments may be made by increasing the frictional resistance in that part of the die through which the clay is moving fastest and decreasing the resistance where the clay is moving too slowly.

Most structural clay products are made with a core in the die to produce hollow parts, such as in the forming of pipes, structural tiles, and cored facing bricks. The cores are placed in the die by means of a bridge at the back of the die, which allows the plastic clay to flow around it. Fig. 47 is a photograph of the back of a die with a bridge holding ten coring fingers that protrude into the die. These fingers can be seen in Fig. 48 from the front end of a die used for forming 10-hole cored bricks. When cores are used in a die, the flow of material around the core bridge and the coring elements must also be balanced in all directions. The first consideration in obtaining a balanced flow is to make sure the cores are centered in the die, and all wall thicknesses of the extruded column are uniform.

Fig. 47. Bridge at the back of the die holding ten coring fingers for cored face bricks. Courtesy of The Bonnot Company

Fig. 48. Extrusion die for a ten-hole cored brick. Courtesy of The Bonnot Company

To reduce the frictional resistance against the sides of the die, lubrication is used. Oils, emulsions, or steam are used for die lubrication by introducing the lubricant under high pressure into an annular groove around the back end of the die. Kerosenelubricating oil mixtures are frequently used. A mixture of 20% kerosene and 80% oil is recommended for clays with low plasticity, and for highly plastic clays the mixture could be 40% kerosene and 60% oil. Wax emulsions, mineral oil and water emulsified with diamide, and emulsions of castor and peanut oils have been used for die lubrication. Die lubrication improves die balance, reduces power required for extrusion, and improves the surface of the extruded column.

The major problem in the auger-extrusion method of forming is the inherent introduction of laminations or inhomogeneous texture in the extruded column. The auger flights are primarily responsible for this, but there are also slip planes created in the spacer and the die [37].

We have learned that clay particles become aligned when pressure is applied to a plastic clay mass. The pressure exerted on the clay by the rotating auger flights orients the clay grains as they push the deaired clay into the die. This orientation puts spiral laminations into the extruded column, consisting of localized concentrations of clay particles all oriented parallel to the auger flights. The new directions of applied forces, as the clay moves through the throat of the die, causes slip planes to develop, which also result in planes of preferred orientation. The pressure on the clay column exerted by the walls of the die causes alignment of clay particles parallel to these walls. This puts an oriented skin on the column, consisting almost entirely of clay-mineral particles. In addition, if there is considerable friction or imbalance as the clay flows through the die, slip planes and fractures can be created in the interior of the product being formed.

If these areas of preferred orientation are not properly dealt with in the extrusion process, they result in laminations that open up as internal cracks during drying and firing of the product. When this kind of internal structure persists into the final product, failures by spalling and disintegration occur in service, especially where freezing temperatures are encountered and where there are soluble salts that tend to crystallize in these open flaws.

It might seem at this point that all is lost in trying to make durable structural clay products by the auger-extrusion method; however, such is not the case, because there are a number of things that can be done to heal these inherent laminations into a coherent body.

The most important procedure in the extrusion process to deal with the problem of laminations, is deairing the pugged clay. When air is present where the extrusion auger is acting to push the clay through the die, the cuts of the auger flights entrap air between the oriented clay planes, and these air pockets prevent the oriented surfaces from coming into intimate contact. As the auger forces the plastic mass into the die, the pressure compresses the enclosed gases, causing the air pockets to become smaller in volume; however, as the compressed column moves out of the die, the pressure is released and the compressed gases expand to open the laminations again. This expansion caused by the release of internal pressure can create additional cracks within the product. All of this action is eliminated by applying an adequate vacuum in the deairing chamber.

In order for the vacuum chamber to do its job properly, a vacuum of 26 in. (66 cm) to 27 in. (69 cm) of Hg is required. A 24 in.-of-Hg vacuum (61 cm) is not sufficient, and 20 in. (51 cm) of Hg is no better than no vacuum at all. It must be understood that the high vacuum does not eliminate the planes of preferred orientation but allows the laminations to be knit together under the extrusion pressure. In fact when the compressible gases are absent, the extrusion pressure becomes greater because of the incompressibility of the clay-water mass.

The importance of a proper vacuum in extrusion cannot be overestimated, and a word of caution should be introduced here. Some factories have been known to lower the vacuum on their machines when producing certain product lines, for the purpose of creating special texturing effects on the surface of the ware. Such procedure should never be resorted to, as history has proven over and over again that durability must never be sacrificed for decorative effects. It would be better to invent an alternative procedure to apply the desired texture to a well-formed column or discontinue the line entirely.

A second consideration for the elimination of the problem of laminations has to do with the raw materials. The texture and filler particles, previously defined in Chapter 3, tend to interfere with the development of continuous layers of oriented clay grains. The particles do not contribute to the orientation when they are more equidimensional in shape, and they disrupt the preferred orientation when the clay grains have to pack around them. The interrupting action of the texture-filler fraction assists the vacuum in knitting together the auger and slip-plane laminations.

Other factors in reducing the disturbing effects of laminations are incorporated into auger design and operation. One of the reasons for the somewhat starved auger during operation is to reduce the orientation of clay planes introduced by the auger flights; consequently, the laminations are smaller and more randomly distributed throughout the extruded column. The propeller at the end of the auger screw helps to chop up the laminations just before the plastic clay moves into the die throat. The proper helix angle on the auger and the correct ratio between the auger feed and the cross-section of the die are other factors which tend to minimize the undesirable effects of laminations.

The water content is another possible variable in the control of laminations. The amount of water necessary for maximum plasticity is conducive to the introduction of layers of preferred particle orientation. This is another case where a reduction in water content is advantageous if there is already sufficient plastic strength.

Usually die lubrication has little or no effect on the development or healing of laminations from the auger, but it does help to reduce slip planes in the column as it passes through the die. The use of an internal lubricant seems to reduce the severity of laminations, but there will be an accompanying loss of plasticity.

Extrusion by a ram instead of an auger is a way of eliminating laminations in the product; however, this must necessarily be an intermittent instead of a continuous process. Since sewer pipe is made by an intermittent process where a length of pipe is extruded, the extrusion stopped, and the pipe cut off before the extrusion is started again, the process is suited to ram extrusion. Large-diameter pipes are often made by vertical ram extrusion. Fig. 49 shows the ram extrusion of a large sewer pipe using a steam-lubricated die. The pug mill feeding tempered clay to the ram cylinder is located above the extrusion die assembly.

Sewer pipes are also made by auger extrusion either horizontally or vertically, and all of the things previously written about plastic extrusion apply equally well to pipes as to bricks and tiles. Fig. 50 shows the horizontal extrusion of a plain-end sewer pipe.

Fig. 49. Ram extrusion of large sewer pipe

Fig. 50. Horizontal extrusion of sewer pipe

The complex shapes in sewer-pipe fittings, such as Y's and T's are put together manually. The lengths of extruded pipe are prepared for fitting together in the plastic state, and in some cases, they are attached and fitted together with plastic clay while in other factories, the pieces are glued together with epoxy cement after firing. The people employed in this part of the forming process are skilled workers in clay or plastics. The production of S-traps is done directly off a vertical extrusion machine by a worker who bends the soft column into the proper shape as it is being extruded. Obviously, this method of forming requires a highly plastic clay column.

4.4.4. Plastic Pressing

Quarry tiles other than square or rectangular, some decorative floor tiles, and some terra cotta pieces are formed from plastic clay by a ram-die process somewhat suggestive of the soft-mud brick process. The clay batch is pugged, deaired and extruded, and the column cut into slugs of size appropriate for the piece to be pressed. The plastic clay slug is placed on the bottom part of the die made of plaster of paris; then the upper plaster die is lowered to shape the piece between the two dies. The press closes to a fixed position to maintain the correct thickness of the piece. Rapid removal of the pressed piece is effected by forcing compressed air through the porous bottom die and holding a vacuum on the porous upper die until the press is open. The final step in forming of these tiles takes place after firing when they are ground to perfect shape and size.

4.4.5. Cutting of Extruded Columns

For brick and structural tile production, the continuously extruded column must be cut in units of the proper size. A common machine for cutting the moving column is the reel-type wire cutter shown in Fig. 51. The extruded column can be seen running

Fig. 51. Reel-type wire cutter for brick production

through the cutter with the cut column emerging to the right. This machine moves back and forth in synchronization with the column speed. The cycle starts in a position with the reel closest to the die, and as the column passes through the cutter, a limit switch activates the horizontal movement of the cutter when the column reaches the end of the machine. It then follows the column at the column speed, and during this travel, the reel rotates spring-loaded cutting wires through the column. The synchronization of the rates of travel of the column and cutter allows the cuts to be made straight and true. With the equipment shown in Fig. 51, twenty-one bricks are cut at a time without stopping the extrusion process. After cutting, the forward motion of the cutter stops, and it returns to its original position until the extruded column again fills the cutting machine.

Although the reel-type cutting process has been adapted to automatic setting of plastic ware on kiln cars, the method is used mostly when the ware is picked off the conveyor belt and stacked on kiln cars by hand labor. In the industry this process is known as "hacking," whether it is done manually or automatically. A photograph of the manual process is presented as Fig. 52. Note that the kiln cars are resting on elevators to make the workmen's job easier and faster.

Fig. 52. Manual hacking of face bricks with kiln cars on elevators

Fig. 53. Column section being wire cut for automatic hacking of large bricks

The newer automatic hacking machines used in brick factories employ a somewhat different method of cutting the extruded column, which is designed to fit the automated process. First, the column is cut roughly into 4- to 6-foot (1.2 to 1.8 m) lengths by a guillotine. Depending on the thickness of the product, the machine sometimes places two of these column sections together before final cutting. The column section is moved to the automatic hacking machine where it is pushed through fixed wires that complete the forming operation. Such a cutting operation is shown in Fig. 53. The scrap pieces at the ends of the rough-cut section will be returned to the pug mill for reuse. The precisely cut bricks are moved immediately to a spacing table where they are arranged for automatic pickup and transfer to kiln cars.

4.4.6. Automatic Hacking of Bricks

The driving force for the development of automatic hacking was the cost and undependability of the semiskilled labor required for manual handling. Careless workmanship caused breakage and kiln wrecks when the stacks were not set properly. The economics of the setting process have been improved by the introduction of these machines; however, there was more flexibility in the hacking of different shapes and in the design of the setting when manual labor was used. Each automatic setting machine has reduced the labor force from about eight semiskilled workmen to one skilled technician.

Automatic hacking machines in operation are shown on Figs. 54 and 55. The plastic, extruded bricks in Fig. 54 have just been placed on a kiln-car deck in one operation by the lifting and transporting mechanism overhead. The inflatable bags that grip the bricks can be seen in the photograph. This machine is setting the hack shown in Fig. 38. Soft-mud bricks, having been dried on pallets, are being automatically set on a kiln car in Fig. 55. Note the different setting pattern which is more appropriate

Fig. 54. Automatic setting of stiff-mud face bricks on a kiln car deck

Fig. 55. Automatic setting of soft-mud face bricks after drying

to this type of brick. The setting head of this machine is holding one course of bricks just previous to adding them to the stack on the car.

Large-diameter sewer pipes were machine-handled long before hacking machines were perfected for face bricks, because the weight made large pieces difficult to move manually. Sewer-pipe handling machinery has been carried over to small pipes in most factories where they are sometimes set horizontally on racks for drying. Large pipes are still set in a vertical position for drying.

4.4.7. Dry-Press Forming

Dry-pressing is really a damp-pressing operation where the clay-containing materials are tempered with 6% to 8% water before forming. Clay materials with this water content are granular rather than plastic. The dampened materials are put through a rather coarse screen to make the granular aggregates of particles fairly uniform. In this state the material can be poured into a press mold like salt. Hydraulic rams are used to compact the damp charge under pressures of from 5000 to 10,000 psi (352 to 703 kg/cm^2). The compressed piece is ejected from the mold by the lower ram and at this time has enough strength to be handled—carefully. Binders are usually added to the mix to add dry strength and to facilitate pressing by internal lubrication.

The preparation of the clay materials for dry pressing floor and wall tiles is more like the conventional whiteware procedures than those of the structural clay products industry. Relatively pure materials are proportioned and mixed in a blunger with excess water. After a thorough blending, the slip is filtered and the clay body dried, pulverized, and dampened with just the right amount of water. The blended materials are fed to the presses for forming when they have been worked through a screen.

Dry-pressing of face bricks was a fairly common practice 70 to 80 years ago in the United States, but it was superseded by the extrusion process, which gave greater production rates. Now it may be time to reconsider the dry-pressing process for the manufacture of face bricks. During this lapse of time, presses have become bigger and faster, and they may be able to produce brick economically now. There are certain advantages to be gained by reintroducing the dry-press method. To start with, the problems and cost of drying would be eliminated, and the dependence of the industry on clay minerals would be overcome. Products could be made more uniform in size and shape because only firing shrinkage would be encountered, and this need not be appreciable if materials other than clays are used.

References

1. Bernal, J. D., and R. H. Fowler: A theory of water and ionic solution with particular reference to hydrogen and hydroxyl ions. J. Chem. Phys. 1, 515–48 (1933).
2. Forslind, E.: A theory of water. Royal Inst. Cement and Mortar, Bull. No. 16, Stockholm, 1951.
3. Linnett, J. W., and A. J. Poe: Directed valency in elements of the first short period. Trans. Faraday Soc. 47, 1033–44 (1951).
4. Runnels, L. K.: Ice. Sci. Am. 215, 118–26 (1966).
5. Morgan, J., and B. E. Warren: X-ray analysis of the structure of water. J. Chem. Phys. 6, 666–73 (1968).
6. Hunt, J. P.: Metal Ions in Aqueous Solution. New York: W. A. Benjamin, Inc. 1963.
7. Weyl, W. A.: Surface structure of water and some of its physical and chemical manifestations. J. Colloid Sci. 6, 389–405 (1951).
8. Hendricks, S. B., and M. F. Jefferson: Structure of kaolin and talc-pyrophyllite hydrates and their bearing on water sorption of clays. Am. Min. 23, 863–75 (1938).
9. Alexander, L. T., and T. M. Shaw: Determination of ice-water relationships by measurement of dielectric constant changes. J. Phys. Chem. 41, 955–60 (1937).
10. Bodman, G. B., and P. R. Day: Freezing points of a group of California soils and their extracted clays. Soil Sci. 55, 225–46 (1943).
11. Grimshaw, R. W.: The Chemistry and Physics of Clays ..., 4th Ed. New York: Wiley-Interscience. 1971.
12. Gouy, G.: Sur la constitution de la charge électrique à la surface d'un électrolyte. Ann. Phys. (Paris) 4, 457–68 (1910).
13. Lawrence, W. G.: Theory of ion exchange and development of charge in kaolinite-water systems. J. Am. Ceram. Soc. 41, 136–40 (1958).
14. van Olphen, H.: An Introduction to Clay Colloid Chemistry. New York: Interscience Publishers. 1963.
15. Button, D. D.: The Effect of Temperature on the Charge of Kaolinite Particles in H_2O Suspensions, Ph. D. Diss., N.Y. State College of Ceramics, Alfred Univ., April, 1963.
16. Button, D. D., and W. G. Lawrence: Effect of temperature on the charge of kaolinite particles in water. J. Am. Ceram. Soc. 47, 503–9 (1964).
17. Grim, R. E.: Clay Mineralogy, 2nd Ed. New York: McGraw-Hill. 1968.

18. Mackenzie, R. C.: Density of water sorbed on montmorillonite. Nature 181, 334 (1958).
19. Macey, H. H.: Clay-water relationships. Proc. Phys. Soc. (London) 52, 625–56 (1940).
20. Grim, R. E.: Some fundamental factors influencing the properties of soil materials. Proc. 2nd Intern. Congr. Soil Mech. 3, 8–12 (1948).
21. Kingery, W. D., and J. Francl: Fundamental study of clay: XIII. Drying behavior and plastic properties. J. Am. Ceram. Soc. 37, 596–602 (1954).
22. Lawrence, W. G.: Plastic Properties, in Clay-Water Systems, W. G. Lawrence, ed. Alfred, N.Y.: Alfred University. 1965.
23. West, R.: The Plastic Behavior of Some Clays, in Clay-Water Systems, W. G. Lawrence, ed. Alfred, N.Y.: Alfred University. 1965.
24. Macey, H. H.: Experiments on plasticity. Trans. Brit. Ceram. Soc. 43, 5–28 (1944).
25. Bloor, E. C.: Plasticity: a critical survey. Trans. Brit. Ceram. Soc. 56, 423–81 (1957).
26. Bloor, E. C.: Plasticity in theory and practice. Trans. Brit. Ceram. Soc. 58, 429–53 (1959).
27. Buessem, W. R., and B. Nagy: The Mechanism of the Deformation of Clay. Nat. Acad. Sci.-Nat. Res. Coun. Pub. 327, Clay and Clay Minerals, pp 480–91, 1954.
28. Kellogg, B. C., and T. J. Sonneville: Rheological Properties of Plastic Clay and Slip with Respect to Flocculation and Deflocculation, B. S. Thesis, N.Y. State College of Ceramics, Alfred University, May, 1974.
29. Astbury, N. F.: A plasticity model. Trans. Brit. Ceram. Soc. 62, 1–18 (1962).
30. Pyle, R. E., and P. R. Jones: The effects of wetting agents on the physical properties of clay bodies. Am. Ceram. Soc. Bull. 31, 233–36 (1952).
31. Robinson, G. C., and J. J. Keilen: The role of water in extrusion and its modification by a surface-active chemical. Am. Ceram. Soc. Bull. 36, 422–30 (1957).
32. Hogue, C. H.: Evaluation and effects of additives in brick making. Am. Ceram. Soc. Bull. 49, 1052–56 (1970).
33. Hodgkinson, H. R.: The shaping and preparation of clay in Germany. J. Brit. Ceram. Soc. 7, 8–12 (1970).
34. Tatnall, R. F.: Globe Brick Co. Achieves automatic pugging. Ceram. Age 78, 27–30 (1962).
35. Connor, J. H.: Mechanism of pugging processes. Am. Ceram. Soc. Bull. 45, 183–86 (1966).
36. Blume, A. J.: Extrusion die design. Am. Ceram. Soc. Bull. 51, 174 (1972).
37. Hodgkinson, H. R.: The mechanics of extrusion. Claycraft 36, 42–48 (1962).

5. Drying Process

5.1. Fundamentals of Drying Clay Bodies

The drying stage of the manufacturing process can be hampered with serious problems if a thorough understanding of the mechanism of drying claywares is not fixed in mind. While it is desired to keep the formed ware as near to its original size and shape as possible, the shrinkage characteristic of clay-water systems on drying can cause distortion, warping, and even cracking if the process is not carried out with a full appreciation of the mechanism. Some kind of energy, conventionally heat, is necessary to bring the moisture out of the clay piece and an efficient use of this energy is a significant economic factor. The rate of drying is also an important consideration. In factory practice there are as many or more losses of products during drying as there are in the firing operation.

That drying of clay products is complicated can be deduced from the long list of factors that affect the process. Some of the factors relate to those already of concern to the forming operations, and others are derived from the application of energy for water removal. All of the factors listed below contribute to shrinkage and the cost of drying, both of which must be controlled for a successful operation:

1. The *initial water content* affects directly the magnitude of shrinkage and the cost of removal.

2. *Particle size of the clay minerals* determines the potential for shrinkage and the amount of water required for forming, and it affects the drying rate.

3. The amount of *nonplastics* in the mineralogical composition and their particle sizes affect shrinkage and drying rate.

4. The *temperature* of drying is related to the rate of water removal and the final equilibrium moisture content of the ware. The application of heat to a damp clay body sets up moisture gradients within the piece that become part of the drying mechanism.

5. The *partial pressure of water vapor* in the drying environment is a factor like temperature in influencing the drying rate. In industrial practice the partial pressures are monitored by measuring relative humidity, which is the percent of water vapor that air contains at any particular temperature based on the amount it can hold when saturated. If the air contains all the water it can hold, it has 100% relative humidity, but an increase in temperature from this condition lowers the relative humidity.

6. The *velocity of air* affects the rate of drying by controlling moisture gradients near the drying surface. This effect is operative up to some optimum velocity.

Drying of clay wares is the reverse of adding water to develop plasticity, and the mechanisms of drying can be related to those of plasticity development. The long flat portion of the plastograph curves in Fig. 34 was described as the tight adsorption of water molecules, followed by construction of water hulls about each clay grain. The molecules involved in these processes will be the last to be removed in drying. Plasticity was developed when a little free water was present; so the removal of this unbound water will be the first stage of drying.

In order to study the fundamental mechanisms of drying, it is necessary to set up conditions for the control and measurement of the variables involved. This we have done at the New York State College of Ceramics by means of a circulating air drying cabinet in which both temperature and relative humidity can be controlled. A spherical, plastic clay specimen is suspended in the controlled atmosphere with calibrated transducers attached to measure weight and dimensional changes. From the experiments made with this equipment, a great deal of fundamental drying information has been obtained [1]. Similar equipment and other arrangements not so similar have been used by other laboratories to add to our total knowledge of drying of clay bodies.

Two drying curves, typical of all clay bodies, are shown in Fig. 56 for specimens made from 50% fine-grained pure kaolin and 50% flint [2]. The curves represent two different drying conditions—one at 50°C and 50% relative humidity and the other at 70°C and 40%. The latter produces faster drying because of the higher energy input and the lower vapor pressure of moisture in the atmosphere. (Curves like these were reported by Funk [1] for whiteware and shale brick bodies.) Curves of this type

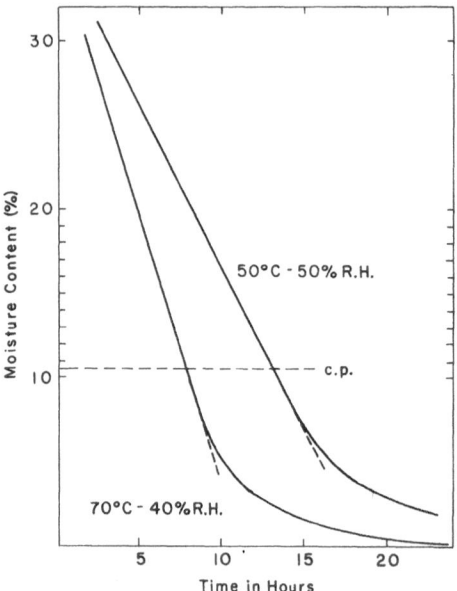

Fig. 56. Typical drying curves for clay bodies. After Thomas [2]

show two mechanisms. Over the first portion of drying time, the moisture contents are linear functions, but in the later stages the curves become exponentially slower.

The linear part of the drying curves represents a constant rate of drying, which means that the rate of evaporation of water from the surface of the piece and the concentration of water on the surface are constant. Such a situation can exist only when water is migrating to the surface from the interior as fast as it is evaporated. This is the free water which is able to move through the capillaries around the various mineral grains to the surface under the pressure of a moisture-concentration gradient. The capillary-flow idea in this stage of drying was adopted by Hougen [3] and his colleagues when they found that diffusion equations could not be applied. It has been found, however, that the rate of evaporation is about half or less that from the surface of pure water, and the exact difference in rate is probably dependent on the mineralogy and the particle-size distribution of the specimen [4, 5]. This migration of water from the interior of a clay piece sets up moisture gradients within it. The gradients have been measured, and it was found that the center of a drying clay body persists in having a higher water content than any point nearer the surface [6].

The movement of water through a clay body during the linear drying period has been followed by use of a radioactive tracer element. By dissolving a small amount of sodium sulfate containing a radioactive isotope of sulfur in the water used to make the clay plastic, the migration of water was followed by radiation counters and photographic film. Fig. 57 shows the build-up of sodium sulfate on the surface of a clay

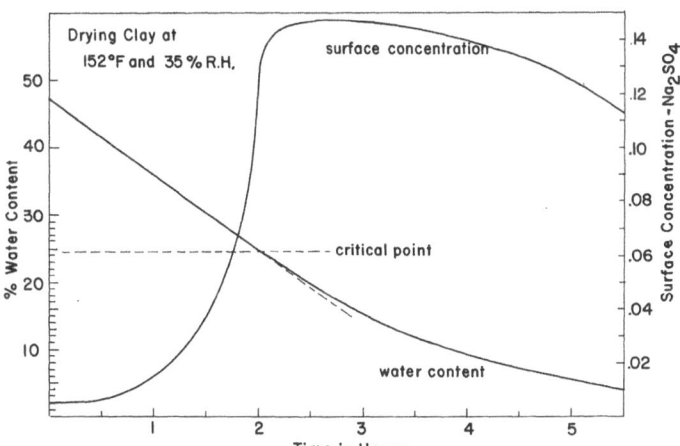

Fig. 57. Increase in surface concentration of a soluble radioactive salt during drying of a clay specimen under constant conditions. Compiled from the work of Klein [5]

specimen in relation to its drying curve. The water continued to bring the radioactive salt to the surface of the specimen during the linear drying period, but at the end of the period the migration stopped abruptly, then receded a short distance into the specimen [5].

Very careful experiments on drying often show some deviation from linearity during this first drying period, but when correction is applied for the shrinkage of the specimen, the constant rate is confirmed [4]. When the shrinkage, and thus the change in drying area of a specimen, are relatively small, the deviation from linearity is not observed, but a correction is necessary as the shrinkage becomes greater.

The change from a linear drying period to an exponentially slower period indicates a change in the drying mechanism. As shown on Fig. 57, the water stops migrating to the surface because the free water is exhausted. Then the water adsorbed on the clay particles, which constitutes the structured water hulls, is removed. This bound water becomes increasingly more difficult to remove as the water hulls become thinner and thinner.

In the exponential drying stage, the adsorbed water is evaporated at its location, and the vapor molecules must find their way out of the piece by way of open channels or capillaries. This diffusion process becomes progressively slower as the evaporation front recedes toward the center of the ware; however, the falling-rate mechanism is more complicated than simply a lengthening of the diffusion distance. The diffusion equation based on moisture gradient, geometry of the specimen, and time does not fit the observed results; and we are forced to presume that the diffusion constant of the equation changes as the adsorbed water hulls become thinner [3]. This means that the water molecules first removed have relatively low bonding energies to the clay, but as evaporation proceeds the remaining water molecules become bonded to the clay ever more tightly, and more energy is required to remove them.

An entirely different approach to the understanding of drying was borrowed from the soil scientists and applied to clay-water systems by Packard [7]. He has worked out the *moisture stress* in clay bodies as a more appropriate parameter for describing the movement of water through clay bodies than the consideration of moisture gradients. Moisture stress is the energy required to remove water from a clay body which, in turn, is equal to the bonding or holding energy of the water to the clay body. It can be expressed in ergs/gram, and a transformed logarithmic function, pF, is commonly used in clay-water systems because of the wide range of values encountered.

The relation between total water content and moisture stress adds credibility to the constantly changing diffusivity theory of the falling-rate period of drying. Moisture moves in clay bodies only in response to a gradient of the potential energy of the water and not necessarily to a moisture concentration gradient. In other words, it is quite possible to have a moisture content differential in a clay piece while the moisture stress is uniform. This explains why Moore [8] and many others have observed water gradients in plastic and partially dried clay masses even at equilibrium which, in itself, is only obtained after very long times. Fig. 58 shows the moisture stress in a typical whiteware body from plastic to dry state [7]. For now, note the shape of the curve from c to d, which is the falling-rate period where we presume the diffusivity is constantly changing. The drastic changes in slope show that water molecules are bonded to the clay with exponentially increasing energy as the moisture content decreases. This explains why Hougen [3] could not use ordinary diffusion equations to express the process of drying in the falling-rate period, because the diffusion constant was constantly changing.

The change from a linear rate to a falling rate is called the *critical point* (c.p.), and it is expressed as the percent of moisture in the body at this point. As can be

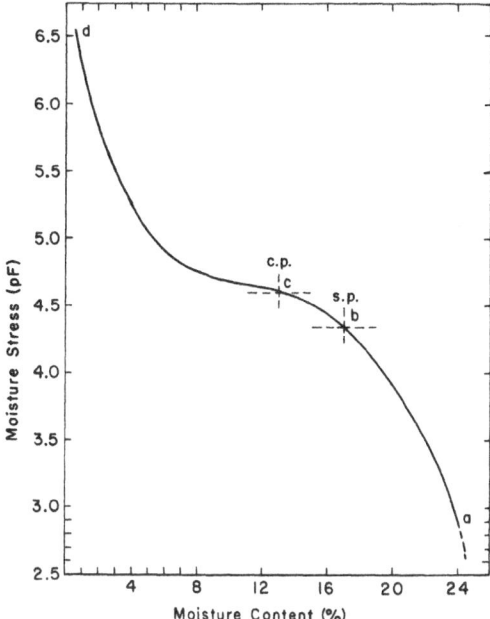

Fig. 58. Moisture stresses in a whiteware body on drying. Data from Packard [7]

seen on Fig. 56, the exact water content of the critical point is difficult to read on the drying curves, but when the drying rate is plotted against water content, the point is more clearly defined, as shown on Fig. 59. The rates were taken from the

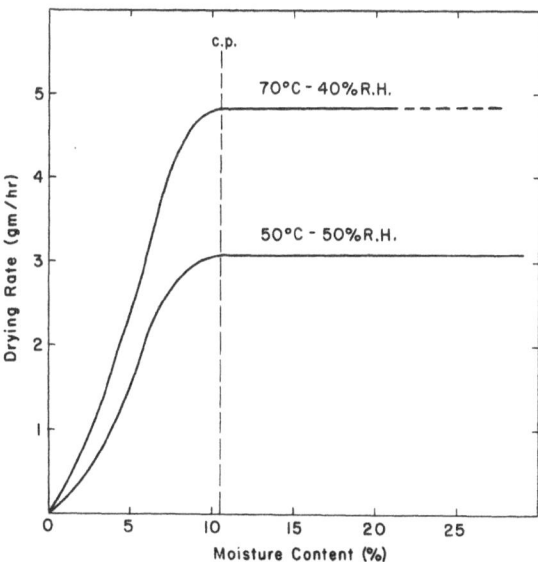

Fig. 59. Drying rates taken from the curves of Fig. 56. After Thomas [2]

curves of Fig. 56, and the dashed line represents that part of the curve where a shrink-age correction was applied.

The critical point has a fundamental significance because it is only dependent on the nature of the material being dried, and it is unaffected by drying conditions. Figs. 56 and 59 demonstrate that the critical point at 10.5% moisture is the same for the same material dried at different rates [2]. The point also has a practical signif-icance in that it denotes the approximate end of shrinkage. The drying rate can be increased drastically as the critical point of the material is reached. It marks the end of liquid water migration to the surface, which is the end of soluble salt build-up on the surface if there happens to be salts in the water of plasticity. Some salts commonly found in clay raw materials are troublesome when they are precipitated on the sur-face of ware during the linear drying stage.

The designation of the critical point (c.p.) on Fig. 58 is approximate, based on experience with similar clay bodies. The position was inferred from its relation to the point where shrinkage ceased [7].

Some typical critical points are listed in Table 12. It can be seen that the amount and fineness of the clay minerals affect the critical point. The more clay that is present, the higher will be the percent of water in the body at this point, and the finer grained clays also give higher critical points. The relationships are reasonable because the total surface area of the clay present determines the volume of water that will be adsorbed by the body.

Table 12. *Typical Critical Points for Clay and Clay Bodies*

Material	Critical Point (%)
Fine-grained clays without filler	20–30
Fine-grained clays with filler	10–18
Coarse-grained clays with filler	6– 8
Kaolin clay with 50% filler	6–10
Kaolin clay with 70% filler	9–12
Pulverized, hard shale (low plasticity)	5– 7
Montmorillonite containing clay + filler	18
Fireclay bodies without filler added	10–12

Another interesting point that can be taken from laboratory drying curves is the water content at equilibrium with the drying conditions. Different clays and clay bodies can be compared at this point, provided drying conditions are constant. A clay body with more or finer clay will retain more moisture at equilibrium than one with opposite characteristics. In addition to being material-dependent, the equilibrium water content is also temperature-dependent. As the temperature of drying is increased, the equilibrium point will be at lower moisture contents, even though the partial pressure of water vapor in the environment remains constant [5, 6].

The actual temperature of a clay piece varies as the drying mechanisms proceed at a constant air temperature, relative humidity, and air velocity. In the beginning, for about an hour or so, the temperature of the clay body increases until it reaches the wet-bulb temperature. Throughout the constant rate period, the cooling effect of surface evaporation maintains the temperature of the ware close to the wet-bulb temperature. At the critical point, the temperature of the piece begins to rise toward the dry-bulb temperature. During the falling-rate period, the temperature of the specimen slowly increases and finally reaches the dry-bulb temperature when drying stops and the clay body is in equilibrium with the drying conditions [9].

Temperature gradients in the drying ware set up moisture gradients that are opposed to the normal convective drying mechanisms. Experiments have shown that if plastic clay is inserted into a nonporous closed tube, to prevent evaporation, and a temperature gradient is set up along the tube, moisture will flow to the cool end [4, 8]. Vassiliou and White [10] performed the experiment on china clay containing 17.8% water (near the critical point). After 1032 hours, the moisture content rose to about 27% at the cool end, and about half way along the tube the moisture was about 0.5%. Similar results were obtained by Plaul [11] in quite a different experiment. Vapor flow and capillary condensation was the explanation for this phenomenon by the authors.

In drying clay wares, air must be circulated across the surfaces where evaporation is taking place because there is a tendency for a moisture gradient to be established in the atmosphere close to the piece. For example, the overall drying air may have a relative humidity (R.H.) of 50%, but close to the drying surfaces the R.H. may increase over a short distance to 90%. Apparently, the diffusion of water vapor away from the drying piece is not rapid. Such a situation retards the drying process, since the drying surfaces see only an atmosphere of 90% R.H. It is, then, necessary to have a movement of air across the surfaces of the ware to flatten the atmospheric moisture gradient.

It has been found that the shape of the ware affects the rate of drying. This is partially due to the way the air flow passes across the drying surfaces. The removal of surface moisture gradients is accomplished most efficiently by impingement of circulating air perpendicular to the surface, in contrast to a glancing or tangential angle. In drying cored bricks, the drying is most rapid when air flows perpendicular to the broad or laying side of the ware. This makes the air flow through the core holes as well as impinging on the broad side, and the flow through the holes promotes a more uniform, as well as faster, drying of the piece [12].

The velocity of the air striking the drying surfaces is an important factor in increasing the drying rate, at least up to some optimum velocity when moisture gradients are not allowed to build up at all. Some experiments in high velocity air impingement used velocities of 400 to 460 ft./min. (2.0 to 2.3 m/sec). When velocities of this magnitude were used, it was absolutely necessary to have the air strike the drying surfaces as uniformly as possible to prevent warping caused by excessive moisture gradients in the piece [13].

Hancock [14] noted that the rate of evaporation from a free-water surface doubles for every 230 ft./min. (1.2 m/sec) of air velocity. It is not certain how this factor applies to a wet clay surface, since the rate of evaporation in this case is much slower than from a free-water surface. He found that the rate of drying from a clay-water surface was twice as great for a perpendicular impact as for a tangential direction. He also emphasized symmetrical drying of the ware as an important economy factor.

After a clay body has been completely dried, there is a tendency for the clay to readsorb moisture from the atmosphere. This readsorption takes place if the ware is allowed to remain for some time under ambient conditions, and problems can arise from such readsorption in a rigid body. The amount of readsorption which takes place is dependent on the relative humidity of the atmosphere. Fig. 60 shows the amount of moisture taken up by a dried piece subjected to the entire range of humidities

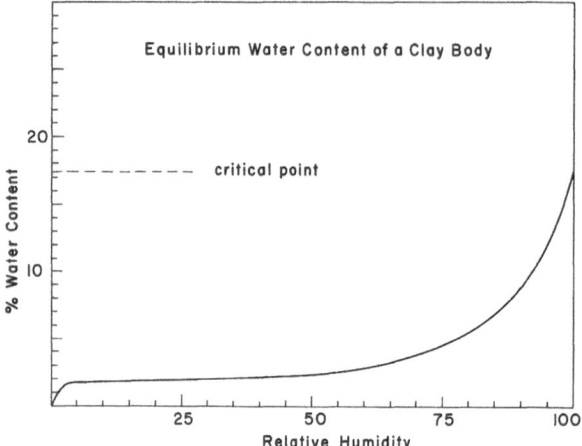

Fig. 60. Equilibrium readsorption of water from the atmosphere by a dried clay body.

until equilibrium is reached. The amount of water adsorbed becomes appreciable above 50% R.H. An interesting characteristic of the critical point of drying comes into play here. If a clay body is held until equilibrium is established in an atmosphere of 100% R.H., it will adsorb the amount of moisture which has been designated previously as the critical point. This behavior further substantiates the theory that water contents up to the critical point represents physically adsorbed water on the clay grains. Curves similar to those in Fig. 60 have been reported by Alviset [6] and Plaul [11].

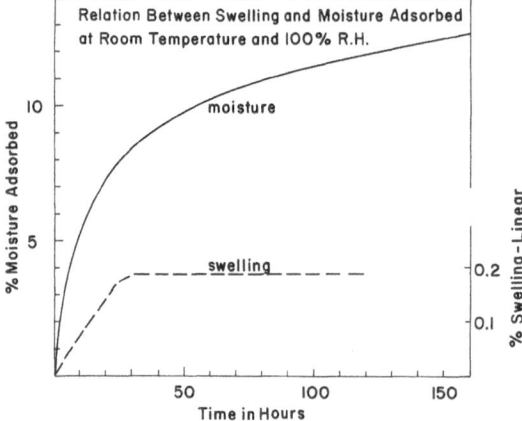

Fig. 61. Swelling of a clay body on readsorption of moisture from the atmosphere

When a dried clay piece is allowed to readsorb moisture, a swelling is often observed, especially with clay bodies containing some montmorillonite. (Recall the expanding nature of this mineral with respect to moisture as described in Chapter 3.) Fig. 61 shows the swelling of a montmorillonite-containing body on readsorption of moisture at room temperature and 100% R.H. The maximum swelling is reached in a relatively short time in relation to the time required for equilibrium. For this body, more than 200 hours were required to reach equilibrium adsorption under these conditions. Readsorption and swelling is an important consideration for the production of clay products because there is an accompanying loss of dry strength.

5.2. Shrinkage, Stresses, and Strength

Shrinkage on drying is of profound concern to the structural clay products industry. Since clay minerals are responsible for shrinkage, the amounts present and their particle sizes determine the shrinkage potential; then, the amount of water present in the plastic clay is proportional to, but not equal to, the shrinkage. These relations can be seen in Fig. 62 where the linear shrinkages for a fine- and a coarse-grained, beneficiated kaolin, both containing 50% flint, were measured while drying at 50°C and 50% R.H. (The fine-clay specimen used for the curve on Fig. 62 is the same as the one used for the plot on Fig. 56.) The initial water content to create good plastic

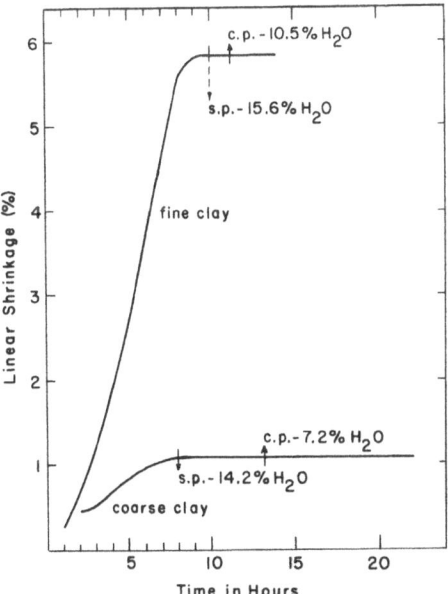

Fig. 62. Shrinkages on drying of a fine-grained and a coarse clay with 50% flint added to each. After Thomas [2]

bodies was about 36% for the fine clay and about 26% for the coarse clay. The shrink-age of the fine-clay specimen is much greater than that of the coarse-clay specimen, and the shrinkage with time is nearly linear under constant drying conditions [2].

The end of the linear drying period is marked on the shrinkage curves of Fig. 62 by the critical point (c.p.), and it will be observed that the shrinkages of these bodies ceased somewhat before the critical points were reached and at higher water contents. Funk [1] has shown that the same relationship existed during the drying of several practical clay bodies. The end of shrinkage on the curves of Fig. 62 has been marked *s.p.*, which designates the *shrinkage limit* point. The shrinkage limit, expressed in percent water content, is reached in most ceramic clay bodies when the water separat-ing the grains has been removed to the point where particle-to-particle contact is achieved throughout the specimen. In other words a skeletal structure is established which prevents the whole body from shrinking more, but localized shrinkage of the fine clay could continue for a while in and around the particles in contact. Perhaps a fine-grained clay with a relatively narrow particle-size distribution and no filler present would have the shrinkage limit coinciding with the critical point, but an experiment to show such a relation is difficult to carry out because of the excessive shrinkage encountered. The presence of a substantial amount of montmorillonite in a clay body would produce a special case where the shrinkage limit might occur a little time after the critical point, but this is only of academic concern.

In support of the particle-to-particle-contact theory as a representation of the conditions at the shrinkage limit, it is notable that when 20% more flint was added to the clays used for Figs. 56 and 62, the shrinkage limit was increased by about 2% water content. This caused the shrinkage limit to occur at 17.2% moisture for the fine clay instead of 15.6% when only 50% flint had been added. The critical point was scarcely affected by the increased flint addition. The addition of 70% flint to the clays instead of 50% flint did not affect the total shrinkage, since the initial water contents were approximately the same [2]. This means that the addition of filler to a clay affects the total shrinkage only modestly, and a large addition, 40% to 50%, is required to reduce shrinkage substantially. Total shrinkage is more related to the water content of the plastic clay at the time of forming than to the amount of filler present. Of course, when a large addition of filler is made, the clay content is reduced and less water is required for maximum plasticity.

Differential shrinkages set up stresses during drying that can cause warping and cracking of the ware. Variations in moisture content in the piece are largely respon-sible for shrinkage differences from one part of the body to another. Such moisture variations can be caused by the drying process, poor mixing of water and clay, pres-sures of the forming operation, and inhomogeneities in mineral distribution [11]. The pressures of forming cause preferred orientation of the clay grains, and this, in turn, promotes moisture differentials. It has been observed that shrinkage is generally greater and more rapid in the direction perpendicular to the planes of orientation than parallel to them [15, 16, 17].

Another unique feature of clays is the increase in strength on drying. As noted in Chapter 4, the shear strength of clay bodies increases from the plastic strength to a maximum when all the adsorbed water is removed. Shortly after the drying of a plastic clay piece begins, it loses its plasticity and becomes brittle, and the modulus of rupture can be measured. The modulus of rupture increases slowly as shrinkage takes place and continues at a low rate of increase until the very last part of the falling-rate period when the modulus of rupture increases rapidly to a maximum value when the piece is totally dry. Fig. 63 shows the importance of removing the last 1% of moisture from an earthenware body. The 10% moisture content represents the *leatherhard* condition [18].

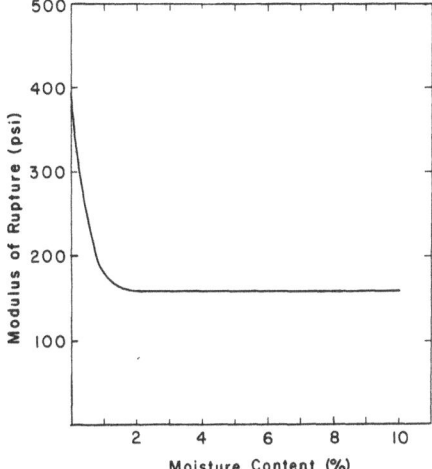

Fig. 63. Increase of dry strength of an earthenware body while drying from the "leatherhard" state to complete dryness. After Holdridge [18]

The raw materials of an earthenware body are not so different from those of a structural clay body except that the particle-size distribution is finer than is usually used for bricks and sewer pipes. The increase in modulus of ruptures of a typical brick and a sewer pipe body on drying are given in Fig. 64. The moisture stresses (pF)

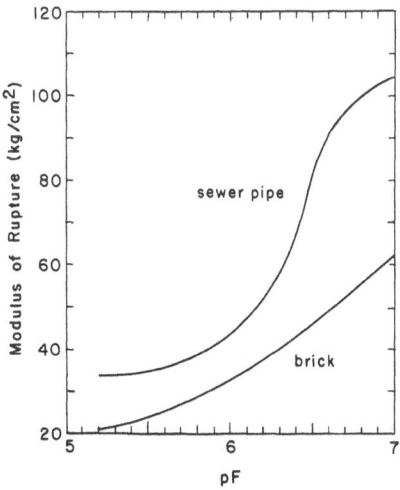

Fig. 64. Increase in dry strength of typical structural clay products in the last stage of drying. After Foster [19]

in these bodies represent about 7% moisture where pF=5.2 and about 0.1% at pF=7. As in the case of the earthenware body, these clay bodies have large increases in strength as the final percent or two of moisture is removed [19]. For the structural clay products industry, this means that if the ware must be handled after drying, it should be completely dried at a temperature of at least 230°F (110°C) for minimum breakage. (To convert the modulus of rupture in Fig. 64 to pounds per square inch, multiply by 14.23.)

Several factors affect the ultimate dry strength. There is little doubt that the strongest clay body would be composed of 100% fine clay-mineral particles [20]; the problem here is to obtain dried ware without warping and cracking. In addition, the dry strength of such a body would be greatest if the clay particles were sodium-saturated. The charge developed on the particles seems to keep them apart during drying, which allows them to move to positions of greatest stability as shrinkage occurs. The addition of nonplastic fillers lowers the plastic strength a little at first, but when fillers reach 60% to 70% of the clay content there is a drastic reduction in dry strength. The method of forming affects dry strength of clay bodies because of the induced preferred orientations of the clay grains and the creation of laminations and slip planes. For example, extruded pieces may be stronger in one direction than another, as was the shrinkage behavior. Deaired clay bodies are stronger than nondeaired bodies because of denser packing and knitting together of the induced laminations [21]. In extruded bodies where laminations and slip planes are common, the dry strength actually increases with the addition of filler which breaks up the continuity of extensive preferred orientations [22].

Readsorption of moisture after drying causes a serious reduction in dry strength, and it should be avoided, especially if montmorillonite is present in the clay body. Clay bodies have been found to lose half of their dry strength by the resorption of water vapor [23]. The loss in strength follows the amount of moisture adsorbed; therefore, strength losses can be inferred from Figs. 60 and 61. The expanding layers of montmorillonite upon adsorption of water causes a drastic loss in dry strength. In the factory, it would be well to avoid letting dried ware stand around many hours in ambient conditions, especially when the humidity is high. The dried ware should be placed in the kiln for firing as soon as possible [24].

5.3. Practical Drying Schedules

The drying of structural clay products is a very specialized activity because there is so much variation in raw materials, methods of forming and handling, and size and shape of products. Each factory requires custom-built dryers and unique drying schedules. It would be impossible to set down specifications for equipment and drying parameters for all of the structural clay products industry. This makes individual reliance on the fundamentals of drying extremely important for design of the most economical process. Fundamentals cannot be ignored, and they can lead the ceramic engineer towards optimum conditions. Above all, he must know the drying behavior of his raw material and such basic constants as the critical point and shrinkage point [25].

An idealized practical drying schedule for clay products is shown on Fig. 65 where the time scale is left unspecified. The drying conditions are shown by curves representing dryer-air temperature and relative humidity. The removal of moisture from clay products, using the conditions specified in Fig. 65, is given on Fig. 66. Note the

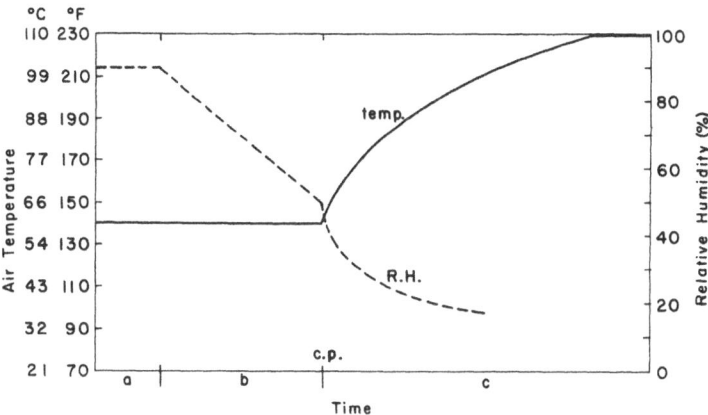

Fig. 65. Ideal practical drying schedule for clay products

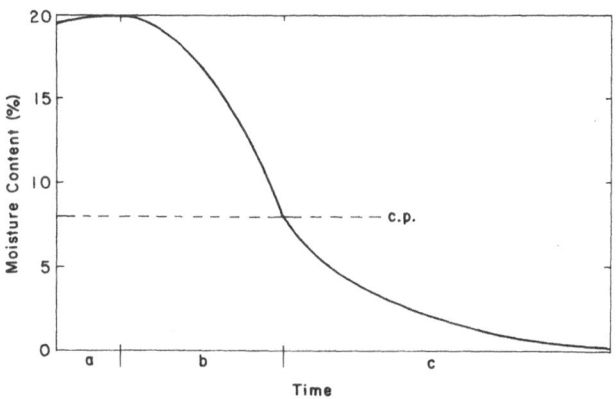

Fig. 66. Drying curve for clay products using the ideal schedule of Fig. 65

slight increase in moisture content in the time period a. The linear drop in relative humidity through period b produces an almost linear increase in the drying rate, and this is reflected in the curve of moisture content down to the critical point. In time period c the exponential increase of the drying power of the atmosphere, as shown on Fig. 65, is designed to complement the exponential, falling-rate period of water removal from the clay body.

The time axes of Figs. 65 and 66 are divided into three parts representing three stages in the factory drying process. The first period a is for ware conditioning where the products are brought up to a predetermined temperature without drying. During this stage, temperature forces moisture toward the center of the clay body, thus creating a moisture gradient. Some condensation of moisture on the surface is inevitable, and a slight expansion probably takes place here, due to the induced moisture gradient and condensation. Thomas [2] found a 0.2 to 0.4% linear expansion of specimens during the conditioning period when the R.H. was 50%. With clay products that are sensitive to drying, it is well to plan for ware conditioning in both the design and operation of the dryer. The linear drying period, where shrinkage takes place, is designated as b on the time axis. To prevent warping and cracking, a gradual drop in humidity, without an increase in temperature, is desirable. About one-third of the total drying time required is devoted to this stage. In actual practice it is sometimes impossible to maintain a constant temperature during period b, and in such cases it would be well to keep a constant vapor-pressure difference between the actual drying atmosphere and that of saturation. This can be achieved by lowering the R.H. more slowly than shown in Fig. 65. After the critical point (more precisely the shrinkage-limit point) is passed, more drying energy can be applied to the ware without danger of warping and cracking. Since this is the falling-rate period, where the removal of moisture becomes progressively more difficult, an exponential increase in temperature and an exponential decrease in relative humidity are in order. This will provide the fastest drying rate possible, but it will still require about two-thirds the total drying time to bring the ware to complete dryness. Some factories extend the dryer temperature considerably above the $230°F$ ($110°C$) level specified on Fig. 65, to preheat the ware before introduction into the kiln. Such practice is appropriate where there is plenty of waste heat and the ware can be transferred to the kiln immediately upon exit from the dryer.

The drying conditions shown in Fig. 65 can be more precisely controlled in a periodic dryer than in a continuous dryer. In a periodic drying operation, the curves become a direct schedule when real times are introduced; but in a continuous tunnel dryer, the curves represent a controlled profile along the dryer from one end to the other, and time is established by the charging rate of cars into the dryer.

Air circulation and recirculation are necessary throughout all phases of the drying process, to provide a uniform distribution of air temperature and relative humidity to

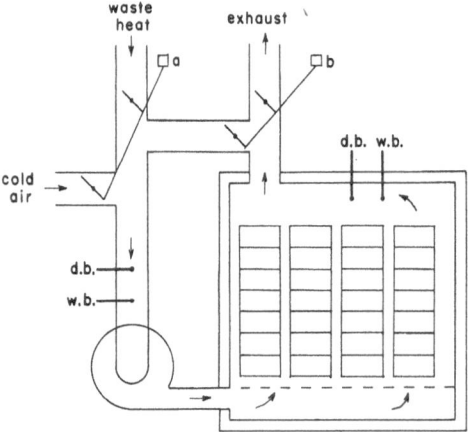

Fig. 67. Schematic drawing of a clay products dryer showing the basic controls

all surfaces of the ware. High air velocities make all stages of drying more efficient, more controllable, and safer from the standpoint of ware losses. The basics of air circulation and temperature-humidity controls on a waste-heat dryer are shown on Fig. 67. The volume of air circulating through the dryer is determined by the capacity of the fan located at bottom left in the drawing. Air temperature to the dryer is regulated at *a* by the dampers on the waste-heat and cold-air supplies. The relative humidity of the air entering the dryer is controlled at *c* through dampers on the recirculation duct and the exhaust stack. Dry-bulb and wet-bulb thermometers to measure relative humidity are shown on the air inlet and in the dryer near the exhaust stack. Additional recirculation for uniformity within the dryer is often practiced, by removing air from the upper portion of the dryer with another fan and reintroducing the air into the lower part. This type of recirculation is designed to effect equal drying of the entire load rather than actually drying one piece after another. The drying of large stacks of ware becomes a real problem in structural clay products plants when one realizes that a single brick can be dried successfully in the laboratory in two to four hours, but the same units require 18 to 30 hours when stacked on cars.

Davies [25] suggests keeping the volume of air to the dryer constant and controlling the temperature differential from inlet to exhaust. The temperature difference required to give a certain drying rate can be calculated. The desired drying rate at any point in time can be calculated from a curve like that in Fig. 66; then the temperature differential can be controlled through dampers like *a* and *b* on Fig. 67.

5.4. Types of Dryers and Energy Sources

5.4.1. Periodic and Continuous Dryers

Besides providing easier control of drying parameters, periodic chamber or periodic tunnel dryers are readily adaptable to pallet drying, where the ware is not stacked in large units. Pallet drying more nearly resembles the laboratory situation, in which each piece is dried separately and large volume elements are not involved. Periodic, chamber dryers are commonly used for sewer pipes because the large units require the precision drying that these dryers can give. Fig. 68 shows chamber dryers in a sewer pipe factory. The doors of the chambers are open because the ware is being set.

Several disadvantages accrue from the use of periodic dryers, and they were responsible for the shift to continuous dryers about the time that continuous tunnel kilns were first introduced to the industry. The big problem with periodic drying is coordination with continuous-flow production. It is a delay point in a continuous process which can only be accommodated by having several units operating in various stages. Energy requirements tend to be greater with periodic dryers because heat is lost each time they are emptied. Preheating such dryers becomes uneconomical and readsorption of moisture between dryer and kiln becomes a real threat to strength and interferes with fast-firing schedules. When several periodic dryers are feeding a continuous kiln, the uniform distribution and use of waste heat is difficult. For this reason,

Fig. 68. Periodic chamber dryers for drying sewer pipes

many periodic dryer installations use a primary source of heat instead of making use
of waste heat from the firing operation.

In spite of the disadvantages of periodic drying, there has been a move back to
chamber drying in the last 8 to 10 years, with new modifications to take advantage
of the faster drying and better control that is possible in a closed system.

Continuous tunnel dryers are integral parts of a continuous-flow process. The
ware is normally set on kiln cars as soon as it is formed, and these cars proceed on a

Fig. 69. Continuous tunnel dryer for production of face bricks

fixed schedule to a tunnel dryer which is completely full. As each car is introduced into the entrance end of the dryer, a car of dried ware comes out the exit end and is immediately placed in a continuous tunnel kiln where the same flow occurs. With this pattern, each time a car of wet ware is put into the dryer a car of finished ware emerges from the kiln. A dryer of this type is shown on Fig. 69. Cars of wet ware are visible in the foreground, about to enter the drying tunnel. The cars of dried bricks come out at the far end of the tunnel and are immediately transferred to the entrance of the kiln, which is located just off the photograph to the left. The waste-heat duct can be seen overhead bringing hot dry air to the dryer. The exhaust stack for the dryer is also visible at the entrance end, where the humidity is highest and the temperature lowest.

A continuous dryer is economical on energy because of the ease with which waste heat from the kiln is utilized and the steady-state conditions of its operation. The difficulties in maintaining steady-state temperatures and relative humidities along the tunnel provide the engineer and designer with interesting challenges. Even with their best efforts, the control over the drying operation cannot be as close as with chamber dryers. Continuous dryers are not as flexible to changes in the size and shape of products as with the periodic type. For this reason, some structural clay products plants employ several short continuous dryers operating on slow schedules to keep each tunnel kiln in continuous operation. With this situation, different drying conditions can be maintained in the several tunnels for the benefit of a variety of products.

A practice which persists in some factories defies control and should be avoided. This is the continuous-periodic operation of tunnel dryers. It is usual in our factories to fabricate products only on an eight-hour day, but the production must be equal to that required for keeping a tunnel kiln operating 24 hours a day, 7 days a week. When wet ware is immediately charged into a continuous tunnel dryer, great pulses of moisture are introduced in the 8-hour working day. Over the weekend, the dryer functions almost like a chamber dryer, yet the temperature profile is that of a continuous operation. This practice should be avoided because the temperature and humidity profiles along the dryer tunnel are constantly changing. It is impossible to control an operation of this type, and it can be only moderately successful when the art of doing it is learned. Even if the products are dried successfully, the whole operation will be inefficient. Continuous dryers should be operated on a regular charging schedule as is a tunnel kiln.

5.4.2. Energy Sources

The use of waste heat from kilns is the most logical source of energy for drying clay wares from the standpoint of fuel conservation, but problems arise when it is attempted which can be solved in the practical use of waste heat. The mechanism for transfer of unused heat from kilns to dryers is by convection where hot air is moved from one process unit to the other by electric fans which also supply the air movement required in dryers. Drying air must be clean and dry, and must have sufficient heat to be able to control the temperature at the dryer. These criteria are not easily achieved in waste heat transfer; however, the industry could do better by recognition of the importance of these requirements and installation of the proper equipment to obtain them.

At this time the only source of waste heat for drying recognized by the industry as suitable is that removed from the cooling ware after firing. Little or no attempt is made to recover waste heat during firing for dryer use, but in tunnel kilns the hot air from the firing operation is used to assist burners in preheating the ware. The problems in its use for drying arise from the combustion of hydrocarbon fuels in the kilns and the evolution of sulfurous gas from many raw materials used in the structural clay products industry. The normal products of combustion are carbon dioxide, water, and sometimes sulfur gases when oil and coal are used. Even though the temperatures may be high, dampness of combustion products makes this air unsuitable for drying at lower temperatures, and we shall see later that sulfurous gases can be extremely troublesome. For these reasons, fresh air is first employed to cool fired ware, and after it is heated to high temperature, it is transferred to the drying operation.

In a factory where periodic kilns are employed, waste heat is taken off only during the cooling part of the cycle. A combination of flues and dampers are installed which allows the products of combustion to be exhausted to the atmosphere during firing; then, after the burners are turned off, the dampers are rearranged to pull fresh air into the kiln and lead the heated air into the dryer duct. At times there is a sulfur contamination problem here, since these gases can be evolved from the cooling ware for some time after firing ceases. This situation requires instrumental monitoring of the exhaust gases for sulfur dioxide, and the kiln is not put on waste heat recovery until no sulfurous gases are present. Some of the tendency to evolve sulfur during cooling can be corrected in the firing operation, and this subject will be discussed in Chapter 6. It would be a much more efficient heat recovery process if the heat generated during firing could also be utilized.

Continuous tunnel kilns also bring fresh air into the cooling zone to assist in cooling the ware and to provide hot air for drying. The furnace-zone exit gases are used to preheat the incoming products. A complex draft situation must exist along the tunnel in order to accomplish both purposes. Fresh air enters the exit end of the kiln, and it is desirable to take off the heated air as close to the furnace zone as possible. At the same time, the main exhaust fan is pulling the furnace-zone gases down the tunnel in the same direction to preheat the incoming ware and to be exhausted to the atmosphere at the charging end of the kiln. This means that a neutral pressure zone must exist between the furnace zone and the take-off point of dryer air. In actual practice this is difficult to balance, and in many cases undesirable furnace gases are pulled into the waste-heat duct leading to the dryer. Evidence of this short-circuiting is the presence of damp air, sulfurous gas, and sometimes carbon soot entering the dryer.

The problems of waste-heat recovery from both periodic and tunnel kilns can be completely solved by installation of heat exchangers. Such units are seldom used in the structural clay products industry and this leads to the question: why not? In the past the greatest barriers were materials and proper operation. In most installations, materials for good transfer of heat are susceptible to corrosion by sulfur trioxide condensation; however, heat exchangers have now been designed for structural clay products plants that are durable and efficient. Temperature controls are used in these heat exchangers to keep all parts above the condensation point of sulfuric acid (338°C or 640°F). The industry should consider seriously the potentials of heat exchangers in several phases of their operation, not the least of which is in drying.

Some dryers in the industry use heat from primary sources instead of waste heat from kilns, and others employ a combination of waste heat augmented by a direct source. Primary heat sources for dryers are generally steam boilers or hot-air furnaces fired with hydrocarbon fuels. Steam heat is used through radiation pipes, with fans for air circulation from the radiators to the clay ware. In no case should products of combustion be allowed to enter the dryer as is sometimes done. Such practice does increase the dryer-air temperature, but most of this advantage is lost by the simultaneous introduction of water vapor. The danger of sulfur contamination is also great. At best the practice is not an efficient use of fuel.

Another form of energy for drying must not be overlooked because of its importance in magnitude. This is the electrical energy put into fans for high-velocity air circulation. As previously pointed out, drying rates can be accelerated by increasing the air velocity passing over the ware without an increase in drying air temperature. It is a way of making more efficient use of heat in the drying process. Some dryers have been built that use high-volume, high-velocity air circulation at relatively low temperature.

Hydrocarbon fuel shortages force us to consider seriously a new energy source for drying clay products that may prove to have great advantages. Dielectric heating has been used successfully in the whiteware industry, and it has now the potential for economic use in the structural clay products industry [26], with the advantage of heating the water in the whole piece at once, thereby discouraging temperature-induced moisture gradients and effecting rapid drying with less stress development. With high-frequency radiation and air circulation, clay pieces could be dried in a few minutes instead of hours as is the current situation. More details of the application of dielectric heating to the drying of clay products will be given in Chapter 12.

5.5. Heat Balance in Dryers

The calculation of a heat balance on a dryer provides many interesting bits of information, especially if the overall equation is put on a computer so that a series of solutions is not tedious. The heat balance which we shall derive here will be for a continuous tunnel dryer where a temperature profile is held constant. The variables that are usually left in the equation for further manipulation are total fan capacity, F, for the circulation of air through the dryer; production in terms of cars of ware per day, C; temperature of ambient air, T_1; temperature of entering air T_3; and the temperature of the air being exhausted from the dryer, T_2. Constant values that seem appropriate can be given to all of these variables except two; then one of these two can be given a series of values and the heat balance solved for the final variable. By this kind of manipulation, one can calculate:

1. the effect of ambient air temperature on the production schedule,
2. the effect of fan capacity on production,
3. the effect of increasing the temperature differential on production or overall drying rate,
4. the necessary fan capacity for a predetermined production schedule.

To make a heat balance, we must add up all of the heat inputs and equate with the sum of all heat uses and losses. For example, a heat balance equation for a continuous tunnel dryer might be set up for an approximate solution as follows, where the waste-heat input equals the sum of the heat uses and losses:

$$FMGdH_a(T_3 - T_2) = W_cH_c(T_3 - T_1)C + W_w(T_2 - T_1)C + W_w\Delta HC$$
$$+ W_rH_r(T_3 - T_1)C + W_sH_s(T_3 - T_1)C$$
$$+ L_h(T_3 - T_1) + L_c(T_2 - T_1)$$
$$+ L_t[(T_3 - T_2)/2 - T_1] \tag{1}$$

where:

F = entrance air fan capacity in ft.^{3}/min
M = minutes/day, i.e., 1440
G = specific gravity of air in the temperature range of dryer, i.e., 0.001
d = density of water in lbs./ft.^{3}, i.e., 62.4
H_a = specific heat of air in BTU/lb./°F, i.e., 0.24
T_3 = temperature of air entering dryer in °F
T_2 = temperature of air leaving dryer as exhaust in °F
W_c = weight of dry clay in lbs./car
H_c = specific heat of clay in BTU/lb./°F, i.e., 0.22
T_1 = temperature of ambient air
C = production in cars/day
W_w = weight of water in clay in lbs./car
ΔH_v = heat of vaporization in dryer temperature range in BTU/lb., i.e., 1007
W_r = weight of refractories in lbs./car
H_r = specific heat of refractories in BTU/lb./°F, i.e., 0.22
W_s = weight of steel in lbs./car
H_s = specific heat of steel in BTU/lb/°F, i.e., 0.11
L_h = heat losses from sides of hot zone in BTU/day/°F
L_c = heat losses from sides of cool zone in BTU/day/°F
L_t = heat losses from top of dryer in BTU/day/°F

(Note that metric units may be used, provided that they are kept consistent throughout; the numerical values must be modified in that case.)

In setting up Eq. (1), several assumptions and approximations were made that did not deviate sufficiently from reality to interfere with the practical use of the solutions. The water of plasticity in the clay products was raised only to the exhaust temperature of the dryer because most of the water is evaporated at a temperature near this value. In calculating average heat losses through the dryer walls and top, the hot zone was considered to have an inside temperature equal to the entrance air temperature, the cool zone temperature equal to the exhaust air temperature, and the average dryer temperature for heat losses through the top.

The heat losses through the walls and top of the dryer are determined by calculating the coefficient of thermal conductivity of the structure in BTU/in./ft.^{2}/°F. These structures may very well be complex in so far as different types of materials in combination are concerned; so the coefficients of thermal conductivity of complex structures are obtained by the following equation:

$$\frac{1}{K_i} = \frac{t_1}{k_1} + \frac{t_2}{k_2} + \frac{t_3}{k_3} + \cdots + \frac{t_n}{k_n} \tag{2}$$

where:

K_i = overall coefficient of thermal conductivity of a particular type of structure,

t = thickness in inches of each material, including air space, making up the wall or roof,

k = coefficient of each material in the structure.

The heat losses per day (24 hours), L_i, through the composite structures are given by

$$L_i = 24K_iA_i\Delta T_i \tag{3}$$

where:

A_i = area exposed to heat losses in square feet,

ΔT_i = temperature differential across the structure in °F.

Some of the terms in Eq. (1) are constants for calculations based on a daily production involving the specific materials mentioned; therefore, it can be modified by substituting the real values in the appropriate places. When this is done and the equation rearranged, it becomes

$$21.56F(T_3 - T_1) = C[0.22W_c(T_3 - T_1) + W_w(T_2 - T_1) + 1007W_w$$
$$+ 0.22W_r(T_3 - T_1) + 0.11W_s(T_3 - T_1)]$$
$$+ L_h(T_3 - T_1) + L_c(T_2 - T_1)$$
$$+ L_t[(T_3 - T_2)/2 - T_1] \tag{4}$$

Although this is an approximate solution to the heat balance, it can be useful, and the data required for its solution are readily available in any factory situation. The precision of the calculation can be improved by entering more precise data for terms that take on special importance in certain cases.

It is interesting to note that the largest use of heat in this whole operation is the heat required to vaporize water after it has been brought up to the temperature of vaporization.

5.6. Scum Development [27]

Scum on structural clay products is an insoluble white stain found on the surfaces of the products as they emerge from the kiln. It is not usually visible in the dry stage, but the ingredients for scum development are there at that time. These scumming materials are formed on the surface of the ware during the drying operation, and this is why the defect is taken up in this chapter.

Scum is caused by the formation of crystals of magnesium and calcium sulfates and barium chloride on the exposed surfaces of the ware in the drying operation. These compounds are brought to the surfaces of the products during the constant-rate period as was the sodium sulfate illustrated on Fig. 57, because the salts are all soluble in the free water of the plastic body. Once they are deposited there by evaporation, they react with the clays to form insoluble silicates. The scumming salts have two physical properties and one chemical property in common which make them potential scum formers. They have high melting points, are soluble in water, and react with clays to form white, refractory silicates and aluminosilicates. Calcium scum as it is seen on the fired product is anorthite ($CaAlSi_3O_8$), magnesium scum is forsterite (Mg_2SiO_4) or enstatite ($MgSiO_3$), and barium scum is barium orthosilicate (Ba_2SiO_4). Sodium sulfate, mentioned in the description of the scum-formation mechanism, cannot produce scum, because it melts at a low temperature and runs into the porous body where it reacts with quartz and the silicates.

The most common source of calcium and magnesium sulfates is the raw clay. Of the two, magnesium sulfate is the more serious scummer because of its greater solubility. When present in small amounts, they can be destroyed by chemicals added to the clay by means of equipment such as that shown in Fig. 29. Barium carbonate is the additive usually used to control scumming caused by these salts, since it reacts with them to form nonscumming products according to the following chemical equations:

$$CaSO_4 + BaCO_3 \rightarrow CaCO_3 + BaSO_4 \tag{5}$$

$$MgSO_4 + BaCO_3 \rightarrow MgCO_3 + BaSO_4 \tag{6}$$

All of these reaction products are too insoluble to migrate to the surface with the free water in the clay bodies; therefore, the scumming tendency is stopped by precipitation as soon as the sulfates dissolve in the pugging water. To determine the amount of additive necessary, a chemical quality-control test is performed on the raw materials before use to determine the quantity of soluble sulfates present. Since barium carbonate is only slightly soluble in water, it is common practice to add twice as much as is theoretically necessary in order to insure an adequate distribution of the chemical throughout the clay mass.

Barium chloride was mentioned also as a scumming salt, but it would not be found as an impurity in the raw materials. It comes into the picture as another additive that may be used to precipitate the sulfates of calcium and magnesium, especially when they are present in relatively large amounts. $BaCl_2$ is much more soluble than $BaCO_3$; so, it is a more reactive additive. The practice is to precipitate the sulfates up to about two-thirds of the theoretical amount with a solution of barium chloride, then use barium carbonate to make the final total correction. This cautious procedure is used because the least amount of excess $BaCl_2$ will cause serious scumming. In heavy correction requirements the combination additive is more economical than barium carbonate alone, even though the chloride is more expensive, because the total amount of barium chemicals added is much less. The equations for the barium chloride reaction with alkaline earth sulfates are

$$CaSO_4 + BaCl_2 \rightarrow BaSO_4 + CaCl_2 \tag{7}$$

$$MgSO_4 + BaCl_2 \rightarrow BaSO_4 + MgCl_2 \tag{8}$$

The chloride products of these reactions are nonscumming, because they have melting points below the final firing temperature of the products. When these salts melt, the liquid is absorbed by the clay body and reaction to form silicates takes place below the surface of the ware. $BaCl_2$ causes scumming because its melting point is too high for this; therefore, it reacts on the surface.

Another mechanism of scum development can take place under a quite special but fairly common set of circumstances. If the raw material contains carbonates of calcium and magnesium, such as calcite, magnesite, or dolomite, they can be converted to sulfates in the dryer by the action of gaseous sulfur oxides in the atmosphere. It is quite common to have sulfurous gases in the dryer atmosphere if waste heat is taken directly from kilns or if fuels containing sulfur are burned directly into the dryer atmosphere. This mechanism presents another good argument for the use of heat exchangers to provide clean dry air for drying. In addition to the presence of alkaline earth carbonates and the contamination of the drying atmosphere with sulfur gases as requirements for scum development, free water must also be present in the clay body at the time of exposure to sulfur fumes. The migration of water to the surface in the linear drying period is still necessary to bring the newly formed sulfates to the surfaces of the ware. So, there are three conditions necessary for this type of scum development in the dryer—calcium and magnesium carbonates present in the raw materials, sulfur oxides in the dryer atmosphere, and free water in the formed pieces.

There is no additive cure for this process of creating sulfates in the ware. Either the ware must be dried to the critical point in clean air or the sulfurous gases must be eliminated from the production dryer entirely. (Even a few parts per million is troublesome.) The latter is a much better solution to the problem because sulfur gases in the dryer can cause other problems, whether moisture is present or not, as will be discussed in Chapter 11.

Visible clues are helpful in determining which mechanism is active in any particular case of scumming. When the scum is due to uncorrected sulfate salts in the raw materials, the salts migrate preferentially to corners, edges, or fine texture protuberances of the ware where drying is fastest. Under close examination of a fired product, preferably with a hand lens, a coating of scum can be seen only on the relief but not on the bottoms of texture scratches and the like. At times, the visible fingerprints of a workman can be seen on a scummed product. The print is preserved, not from contamination by dirty fingers, but from scum on the fine relief imposed by the finger lines and by the slight pressure bringing extra water with salts to the surface. The differential shading causes the fingerprints to stand out visibly.

When the scum on the surface of fired ware is caused by reaction with sulfurous gases in the atmosphere, there will be a uniform coating of white stain over all surfaces, with no preference to corners, edges, or relief surfaces. A hand lens will show the bottoms of scratches and textures to be as well coated as the elevated portions. Such a coating of scum may obscure the characteristics of uncorrected raw materials when both mechanisms are operating. This situation must always be looked for, since it is discouraging to find scum still present, after going to great lengths to keep sulfur gases from the wet ware. It could very well be that more barium carbonate is called for after the sulfur contamination mechanism is interrupted. This you can tell by close examination.

References

1. Funk, J. E.: Simultaneous weight loss and shrinkage of clays. Am. Ceram. Soc. Bull. **53**, 450–52 (1974).
2. Thomas, R. J.: Application of the Diffusion Equation to Clay Drying Below the Critical Point, M.S. Thesis, New York State College of Ceramics, Alfred University, May 1975.
3. Hougen, O. A., H. J. McCauley, and W. R. Marshall Jr.: Limitations of diffusion equations in drying. Trans. Am. Inst. Chem. Engrs. **36**, 183–209 (1940).
4. Norton, F. H.: Fine Ceramics, pp. 157–77. New York: McGraw-Hill. 1970.
5. Klein, J. D.: Contributions to the Theory of Drying of Clay Bodies, B.S. Thesis, New York State College of Ceramics, Alfred University, June 1954.
6. Alviset, L.: L'aptitude au séchage des argiles et des pâtes—phénomènes de diffusion de l'eau. Ind. Ceram. 524–35 (1967).
7. Packard, R. Q.: Moisture stress in unfired ceramic clay bodies. J. Am. Ceram. Soc. **50**, 223–29 (1967).
8. Moore, F.: The mechanism of moisture movement in clays with particular reference to drying: a concise review. Trans. Brit. Ceram. Soc. **60**, 517–39 (1961).
9. Hursh, R. K.: The drying of clay products. Calif. J. Mines Geol. **52**, 177–91 (1956).
10. Vassiliou, B., and J. White: Vapour pressure-capillarity relationships in clays and their application to certain aspects of drying. Trans. Brit. Ceram. Soc. **47**, 351–78 (1948).
11. Plaul, T.: Das Zustandekommen, Messen und Deuten von Potentialdifferenzen in trocknenden Tonen. Ber. Deut. Keram. Ges. **43**, 547–53 (1966).
12. Balint, P.: Drying experiments with green ceramic ware. Epitoanyag **25**, 265–68 (1973).
13. Bird, G. W., and A. J. Dale: Jet drying of whiteware. Trans. Brit. Ceram. Soc. **51**, 559–72 (1952).
14. Hancock, W.: Drying tableware and other ceramic goods by jet-drying method. Ceramics (London) **5**, 408–10 (1954).
15. Cox, R. W., and W. O. Williamson: Differential shrinkage of clays and bodies caused by particle orientation and its significance in testing procedure. Trans. Brit. Ceram. Soc. **57**, 85–101 (1958).
16. Kilgore, R. V., and W. O. Williamson: Anomalous differential drying shrinkage of clay-quartz mixtures. J. Am. Ceram. Soc. **51**, 181 (1968).
17. Holdridge, D. A.: Effect of moisture content on the strength of unfired ceramic bodies. Trans. Brit. Ceram. Soc. **51**, 401–8 (1952).
18. Foster, P. K.: Moisture stress and the dry strength of ceramic clays. New Zealand J. Sci. **12**, 553–63 (1969).
19. Hofmann, V., and A. Rothe: Plastizität und Trockenbiegefestigkeit von Kaolinen und Tonen ohne und mit Zusatz von Quarz. Ber. Deut. Keram. Ges. **47**, 296–99 (1970).
20. Williamson, W. O.: Strength of dried clay-review. Am. Ceram. Soc. Bull. **50**, 620–25 (1971).
21. Kennard, F. L., III, and W. O. Williamson: Transverse strength of ball clay. Am. Ceram. Soc. Bull. **50**, 745–48 (1971).
22. Pask, J. A.: Measurement of dry strength of clay bodies. J. Am. Ceram. Soc. **36**, 313–18 (1953).
23. Anwyl, R. H.: Deterioration of dried brick. Am. Ceram. Soc. Bull. **40**, 359–61 (1961).
24. Murray, M. J., and E. Tauber: Application of drying theory to design of dryers for the heavy clay industry. J. Austral. Ceram. Soc. **8**, 57–61 (1972).

25. Davies, T. E.: The psychrometry of drying ceramic materials. Ceram. Age **84**, 32–33 (1968).

26. Blin, C., and M. Guerga: Trocknen von Keramikerzeugnissen durch dielektrische Verluste. Keram. Z. **21**, 157–59 (1969).

27. Brownell, W. E.: Scum and Its Development on Structural Clay Products, Structural Clay Products Research Foundation, Rept. No. 4, McLean, Va. 1955.

6. Firing Process

6.1. High-Temperature Reactions in Disilicate Minerals

The object of firing a clay product is to convert a fairly loosely compacted blend of various minerals into a strong, hard, and stable product. In the path of achieving this conversion, many chemical and physical processes come into play. The properties of the final product such as strength, porosity, stability against the action of moisture and chemicals, thermal expansion, thermal conductivity, and hardness are determined by the kind and amounts of the various phases resulting from the firing process.

In the beginning it is appropriate to review the reactions that take place on heating some of the important disilicate minerals. The raw materials for structural clay products contain from 40% to 80% of these minerals; therefore, they are bound to have important influences on the final product and on the way that the final phases are achieved. It will be found that a single disilicate mineral may break down on heating to provide as many as four new compounds, to serve as reactants with each other and other minerals present. In many cases one or more of these decomposition products remains in the final product and exerts its influence on the properties.

The most common tool used in observing high temperature reactions is Differential Thermal Analysis (DTA). It measures the exothermic and endothermic effects that occur on heating a sample at a constant rate, usually around $10°C$ per minute. These heat effects are noticeable on DTA when reactions take place fairly rapidly. Sluggish reactions are not visible, since the temperature differentials caused by them are too small and extend over time periods that are too long. DTA tells us the temperature ranges in which changes are taking place, and from its thermal effects, we can infer the type of reactions occurring. For example, decompositions and crystal transitions are endothermic, but oxidations and reactions to form new phases are exothermic.

More detailed probes into the identity of the phase changes indicated by DTA are made by Thermal Gravimetric Analysis (TGA) and X-ray diffraction. TGA also is a dynamic process achieved by heating the sample at a constant rate. It gives weight changes that supplement the information gained from DTA. If an endothermic peak on DTA is shown by TGA to be accompanied by a weight loss, the reaction is unquestionably a decomposition evolving a gaseous phase. If the reactants and products involved with DTA peaks are crystalline, direct identifications of them are possible by X-ray powder diffraction methods. Samples for X-ray examination can be obtained by quenching from a DTA furnace both below and above the temperature of visible peaks, or other samples can be heated to equilibrium at appropriate temperatures. The actual reaction temperatures are not often easy to determine from the DTA because they are the points where the peaks start. The temperatures of the maxima and minima on the DTA peaks have no physical meaning because of the dynamic nature of the heat treatment. They are simply reasonably reproducible points on the temperature scale. In some cases, but not all, the tops of the peaks may express the end of the reaction; then the rest of the peak is a cooling or heating curve bringing the sample back to the furnace temperature.

Since kaolinite is the purest and has the simplest crystal structure of all clay minerals, its thermal reactions have received considerable attention. A typical DTA curve for a purified, highly crystalline kaolinite is plotted on Fig. 70. Temperature of the furnace is plotted against the millivolts produced by the differential thermocouple.

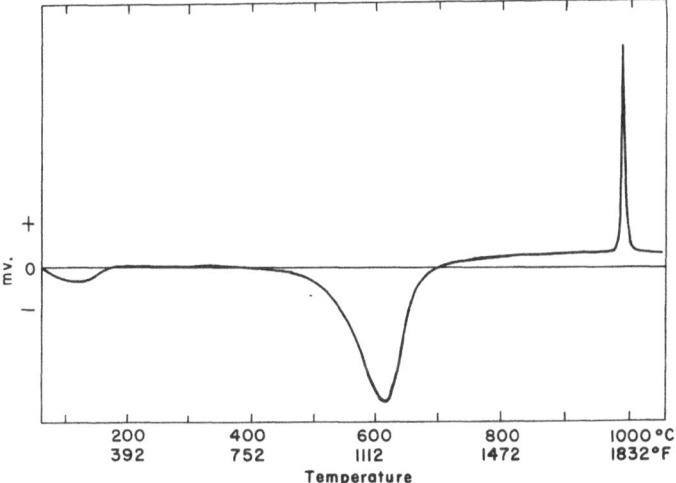

Fig. 70. A typical differential thermal analysis of well-crystallized kaolinite

Quite arbitrarily, the plus sign indicates heat evolved from the sample, i.e., an exothermic reaction, and the minus sign connotes the absorption of heat by the sample as in endothermic reactions. The significant thermal effects for kaolinite up to 1000°C are: an endothermic reaction near 600°C (1112°F) and a sharp exothermic peak around 1000°C (1832°F).

On heating kaolinite no change occurs until about 470°C (878°F) where the OH$^-$ ions of the crystal lattice are expelled in the form of water vapor. The actual temperature of this decomposition and the rate of evolution of moisture is dependent on the vapor pressure of water in the atmosphere surrounding the sample [1]. As the lattice water is driven off, the crystallinity disappears, but a considerable order remains in the a and b crystallographic directions [2]. The hexagonal form of the kaolinite crystals persists through this loss of crystallinity, and can be seen by electron microscopy even after heating to very high temperatures.

Metakaolin is the name given to the amorphous, metastable phase of kaolinite after dehydroxylation, and it has several interesting characteristics. The specific surface increases dramatically; surface free energy for adsorption and catalytic activity increases; and it is highly reactive with other oxides in this form. When metakaolin is formed, the kaolinite layers collapse in the c direction from 7.14 to about 6.3 Å, some of the Al^{3+} ions go into tetrahedral coordination with oxygen, and the tetrahedral silica sheet becomes quite distorted; however, the aluminum and silicon ions do not migrate very far from their original positions in the kaolinite structure [2, 3, 4].

On continued heating, the coordinated Al^{3+} and Si^{4+} ions gain increasing mobility, so that by 855°C (1571°F) a simple rotational movement produces the start of development of a cubic, spinel-like crystalline phase. At this low temperature, the crystallites are visible only by electron diffraction, but the cubic phase is well enough developed at 925°C (1697°F) or a little higher, to be detected by X-ray diffraction [5, 6]. The alumina-silica spinel, previously mistaken for gamma alumina, has an approximate composition of $2Al_2O_3 \cdot 3SiO_2$ and the rest of the silica not included in this rearrangement remains amorphous [6]. This is the structural state up to the sharp exothermic reaction at about 970°C (1778°F).

The exothermic peak on Fig. 70 is the nucleation of mullite from the spinel phase. Such a sudden release of energy must be due to a simple transition of a metastable state to a stable one [5]. Any reaction or transition requiring extensive rearrangement of ions is characteristically sluggish and such an exothermic effect would not be visible by DTA. If one builds a model of kaolinite like that in Fig. 16, then makes a correction for the removal of hydroxyl ions by allowing some Al^{3+} ions to go into 4-fold coordination, and distorts and breaks up the silica sheet, mullite is formed by simple, rotational movements of the coordinated aluminum and silicon ions. As mullite forms, some silicon ions diffuse to grain boundaries. The initial mullite crystals probably have a composition close to $Al_2O_3 \cdot SiO_2$, but as heating continues up to about 1300°C (2372°F), mullite continues to exsolve silica until it finally approaches a stable $3Al_2O_3 \cdot 2SiO_2$ composition [5].

To make a model of mullite from kaolinite, as described above, it will be found that three kaolinite layers are required for each mullite unit cell. A view down the c-axis of a mullite unit cell is shown in Fig. 71A. We are looking down at the ends of

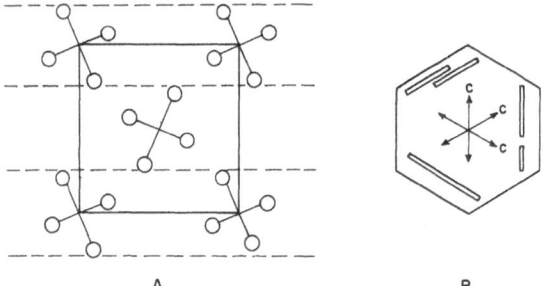

Fig. 71. Relation between mullite development and the structure of the original kaolinite

five chains of aluminum octahedra, each sharing an edge with its two neighbors. The dashed lines represent edge boundaries of the three layers of kaolinite required. The mullite crystal is completed by joining the alumina chains with alternating silica and alumina tetrahedra. Actually, the crystal described is sillimanite, and mullite is the result of silica vacancies along the chains [6, 7].

The c-axis of elongated mullite crystals grow perpendicular to the original c-axis of the kaolinite crystal, and this picture is presented in a plan view in Fig. 71B. The hexagonal outline is that of a kaolinite crystal which has been converted to mullite. Elongated mullite crystals are shown as they grow inside the kaolinite crystal preferentially oriented to the sides of the kaolinite hexagon. The three c directions illustrated are those preferred by the mullite crystals growing in the ab plane of well-crystallized kaolinite [8, 9].

The superfluous silica and that expelled during the development of mullite finally crystallize as cristobalite around 1250°C (2282°F), and the crystallization is visible as a small exothermic peak on high-temperature DTA curves [10]. The first cristobalite detected by X-ray diffraction at this temperature is not well crystallized, but the pattern becomes clearer on heating to 1425°C (2597°F) [6]. The slow development of well-crystallized cristobalite from amorphous silica was observed by Verduch [11] when he found that cristobalite crystals formed at temperatures below 1425°C were quite defective.

The overall chemical equation for the stable phases involved in the heating of kaolinite is

$$3[Al_2(Si_2O_5)(OH)_4] \rightarrow \underset{\text{mullite}}{3Al_2O_3 \cdot 2SiO_2} + \underset{\text{cristobalite}}{4SiO_2} + \underset{\text{water}}{6H_2O}$$
$$\underset{\text{kaolinite}}{}$$

When kaolinitic clays are used for structural clay products, the presence of mullite in the product denotes a well-fired body. Mullite is a hard, chemical resistant phase in clay products, and its elongated crystal structure provides great strength. An underfired product with metakaolin or even poorly crystallized mullite would be totally unservicealbe. As we shall see later, cristobalite is seldom observed as a phase in structural clay products, and mullite is not always a part of products prepared from clays containing kaolinite.

As might be expected from examination of the formula for illite in Table 2, the paths of reactions are quite different from those of kaolinite, and different phases are formed on the way toward a stable product. Surprisingly, in the end the only crystalline phase is still mullite. In this respect, the final phases resemble those of kaolinite, but in other ways they are not the same at all.

Since illite has a somewhat variable composition due to substitutions of one ion for another, the best we can do is describe the course of events that take place on heating a particular illite. This approach will not lead us far astray, because the compositional range of illites is small enough so that deviations in phase development on heating are negligible. Great care was exercised by Phillips [12] in purifying an illitic clay before studying its thermal behavior. Other minerals and carbonaceous matter were removed from the natural sample, and the formula for the pure illite was

$$K_{0.94}Na_{0.05}(Mg_{.54}Fe^{2+}_{0.67}Al_{3.04})(Al_{1.28}Si_{6.72}O_{20})(OH)_4$$

The DTA for this illite clay mineral is given in Fig. 72, and the thermal reactions shown indicate the approximate temperatures where these reactions occur. Equilibrium heating studies tell us only physically adsorbed water molecules are driven off up to 400°C, and this is the cause of the endothermic peak near 130°C. Between 400 and 550°C (1022°F) most of the lattice water is expelled, which accounts for the endothermic peak around 540°C, but TGA shows that water is continually driven off very

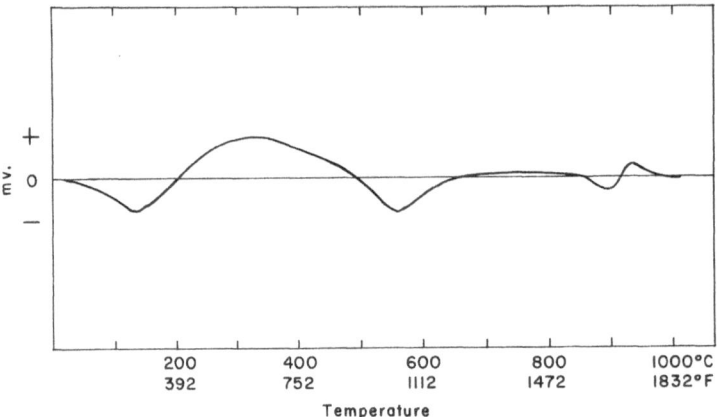

Fig. 72. Differential thermal analysis of a pure illite sample. After Phillips [12]

gradually until melting starts. Unlike kaolinite, the X-ray crystallinity is not lost as a result of the dehydroxylation, but there is an increase in specific surface area. The only change in the X-ray pattern from 550 to 850°C (1562°F) is a drift of the 060 line from 1.50 to 1.53 Å. Between 800 and 850°C a rapid consolidation of the illite structure occurs and is seen by a drastic loss in surface area. This is probably a collapse of the lattice with some distortion, but the crystalline arrangement of the ions still persists. The consolidation process may cause the endothermic effect near 900°C (1652°F).

Above 850°C several reactions take place, most of which relate to the octahedral sheets in the illite layers. The exothermic peak a little above 900°C is caused by the formation of spinel ($MgAl_2O_4$), and it is observed on X-ray patterns of samples heated to approximately 850°C. When spinel first appears all of the illite pattern disappears except the basal spacings 00l, but these also vanish at 900°C when hematite (Fe_2O_3) forms along with spinel. Someplace along the line the ferrous ions of the original illite structure are oxidized to the ferric state. Further heating to about 950°C (1742°F) causes corundum ($a\text{-}Al_2O_3$) to crystallize; then we have three crystalline phases, spinel, hematite, and corundum, present—but no silicates [12].

The first trace of melting on heating this illite occurs around 1050°C (1922°F) and the amount of liquid formed gradually increases until no crystals are left at about 1450°C (2642°F). As soon as melting starts, the initial three crystalline phases dissolve in the fusion and react with silica to form mullite. The amount of mullite increases as the sample is heated to 1200°C (2192°F); then, it begins to dissolve in the liquid [13].

The beginning of melting around 1050°C and the appearance of mullite explains why many structural clay products made from predominantly illitic clays are fired from 1038°C (1900°F) to 1082°C (1980°F). The glassy phase resulting from the fusion promotes a strong body with a reduction in porosity, and mullite gives it strength and durability. Products showing the spinel phase are underfired as far as a good, stable product is concerned.

The high-temperature reactions of montmorillonite are similar to those of illite except that the DTA records show a larger and sometimes double endothermic peak in the temperature range of 150°C (302°F) to 260°C (500°F). This heat effect is attributed to the evolution of interlayer water and the water associated with the hydration of the interlayer exchangeable cations. Even though the lattice water is driven off around 500°C the X-ray crystallinity remains to about 800°C when spinel and hematite forms. As with illite, when melting starts around 1050°C, spinel and hematite quickly dissolves and mullite crystals emerge; so that by 1300°C (2372°F), the only two phases present are mullite and a silicate liquid [14].

The most important nonclay disilicate mineral affecting the properties of fired structural clay products is chlorite. As mentioned in Chapt. 2, it is commonly found in illitic shales and clay deposits derived from them. Because of the alternate stacking of trioctahedral mica and brucite layers, the different structural relations between magnesia and silica produce unique phases on firing.

One of the common chlorite minerals, prochlorite, has been examined carefully on heating to 1000°C [15]. Its structural formula is

$$(Mg_{4.2}Fe^{2+}_{0.78}Al_{1.03})(Al_{1.16}Si_{2.84}O_{10})(OH)_8,$$

and an interesting thing to note here is the absence of alkali ions. The DTA curve for this mineral is found in Fig. 73. No reactions occur up to 500°C, but the peak at 550°C (1022°F) is actually an exothermic reaction and only appears as a return to base line because of a base-line drift during the heating of this material. The heat evolved is caused by the oxidation of ferrous ions in the lattice to the ferric state,

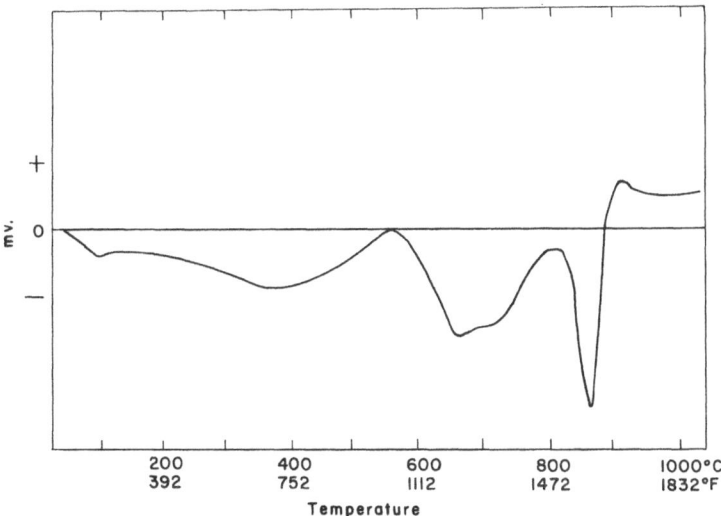

Fig. 73. Differential thermal analysis of prochlorite. After Kellogg [15]

and hematite appears on the X-ray patterns on heating the sample to $550°C$. Between 550 and $720°C$ ($1328°F$), the water is expelled from the brucite layers causing the endothermic effect around $700°C$. During this period, the crystallinity of the prochlorite weakens. At $780°C$ ($1436°F$) the removal of the lattice water from the biotite-like layers is responsible for the endothermic effect on the DTA at about $860°C$ ($1580°F$). Immediately following this dehydroxylation, the chlorite crystallinity disappears, and an olivine crystalline phase, like forsterite (Mg_2SiO_4), appears along with ($MgAl_2O_4$) spinel. The reactions to form these compounds are visible on the DTA curve as a small exothermic peak at $920°C$ ($1688°F$).

Apparently, the magnesium ions which belonged to the brucite layers react with the silica sheets of the adjoining biotite layers, since this is the first 3-layer mineral described that develops a silicate compound before mullitization. As prochlorite is heated above $920°C$, further magnesia-silica reactions take place and clinoenstatite ($MgSiO_3$) appears. At $1000°C$, the crystalline phases present are forsterite, spinel, and clinoenstatite. Possibly these crystals have taken iron into solid solution, because hematite is not present above $780°C$.

Melting occurs in chlorites somewhat above $1000°C$, perhaps at a higher temperature than with the 3-layer clay minerals due to the lack of interlayer alkali ions; however, one can expect from the crystalline phases developed that very fluid melts are going to appear when melting occurs. In practice this is noticed with clay raw materials containing chlorite as narrow temperature ranges for the development of suitable properties in the products.

One might expect that cordierite, (Mg_2Al_3)($AlSi_5O_{18}$), would be developed on heating chlorite, but in the structural clay products industry, the firing temperatures are not high enough when materials are used that contain chlorite naturally. Chlorites have been used to produce a cordierite body by fortifying the composition with kaolinite and firing to a relatively high temperature [16]. This idea may be of some interest to those structural clay products plants having access to fireclays. Cordierite products have a good resistance to thermal shock and are often classed as a low-temperature refractory.

Pyrophyllite is a micaceous mineral which can be used in ceramic bodies to provide alumina and silica without increasing plasticity and shrinkage as would be the case with kaolin. It may occur as a minor constituent in some illitic shales, but it is added directly to floor and wall tile bodies. On heating, pyrophyllite looses its lattice water around $600°C$, and the dehydroxylation is followed by an abrupt expansion close to $800°C$. The crystallinity is lost by the time the mineral reaches $1000°C$, but mullite does not develop until $1400°C$ ($2552°F$). Pyrophyllite is thus a very refractory ingredient to add to ceramic bodies [17].

6.2. Reactions in Typical Clay Bodies

Structural clay products are not made from clay minerals alone as was pointed out in Table 4. Plastic clays require fillers which are sometimes inert in the firing process while others must be classified as reactive. To make a dense, strong body, fluxes are also required. The 3-layer clay and micaceous minerals are, in themselves, fluxes, but kaolinitic clays require additional minerals that will produce some melting at the firing temperatures; however, the amount of liquid phase developed in structural clay products must remain small, probably around 2% or less. Excessive fluxing allows for plastic flow and serious distortion of the ware under load. On the other hand, extensive fluxing is not required to produce optimum properties in structural clay products. On firing, the silicate melts are viscous, and they produce glassy phases on cooling.

In the classical clay body, reactions occur between the clay minerals, fluxing minerals, and minerals that can be called reactive fillers. In order to understand and be able to predict the final properties of a clay product, it is necessary to examine all of the high-temperature reactions that take place, since the phases formed determine the properties to a large extent. The paths of these reactions are determined primarily by the assemblage of minerals in the raw materials; therefore, it is absolutely necessary to know what the starting minerals are and what happens to them on firing.

A basic principle in the theory of firing of structural clay products is that the final development of an acceptable product is a dynamic process [18]. That is, the reactions, both physical and chemical, proceed toward some theoretical equilibrium but never attain it. These reactions are allowed to proceed until the product has developed the phases necessary to give the desired product; then they are arrested by cooling. The idea of a dynamic process presents the feeling of motion, and the rate of this movement becomes important to us if we are to stop it at precisely the right time. If we know where the reactions are going and how fast they are getting there, then, we can tell when the firing process gives an optimum product. The processes occurring in clay products are solid-state reactions, solid-liquid reactions, solid-gas reactions, and sintering which results in larger particles by grain growth.

Structural clay products raw materials are complex mixtures of natural and sometimes synthetic minerals. Because of this complexity, the overall chemical analysis does not tell very much about what is going to happen on firing. A mineralogical analysis is a great deal more valuable. In spite of this apparent complexity, we are fortunate that nature loves simplicity. Pauling [19] stated that the number of essentially different kinds of constituents in a crystal tends to be small which means that even though we have a very complex chemical system, the phases that develop on heating will tend to be simple. For example, a mixture of CaO, MgO, SiO_2, Al_2O_3, and Fe_2O_3 can produce the following simple phases: $CaSiO_3$, $Al_6Si_2O_{13}$, Mg_2SiO_4, and Fe_2O_3. Because of this tendency toward simplicity, two- and three-component phase diagrams are invaluable sources of information, even though they are derived from equilibrium conditions. Phase diagrams are useful to predict the phases that will be developed because they give the ultimate goals of the reactions that are being promoted by firing. These goals may not be achieved during firing of clay products, but we will know where the reactions are going. In other words, the very complex system can be broken down into three or four simpler systems in our high-temperature reaction studies.

Because fireclays, like the one listed in Table 4B, occur naturally as well-balanced clay bodies, they are often used directly for structural clay products. This particular fireclay contains 45% clay minerals, 52% inert filler, 1% reactive filler, and 2% carbon burnout. The clay content is composed of 35% kaolinite and 10% illite, and such a combination puts it into the low-grade fireclay class. Fireclays become higher in grade as the amount of illite decreases, and those with no illite are classed as high-grade and are used primarily for refractories.

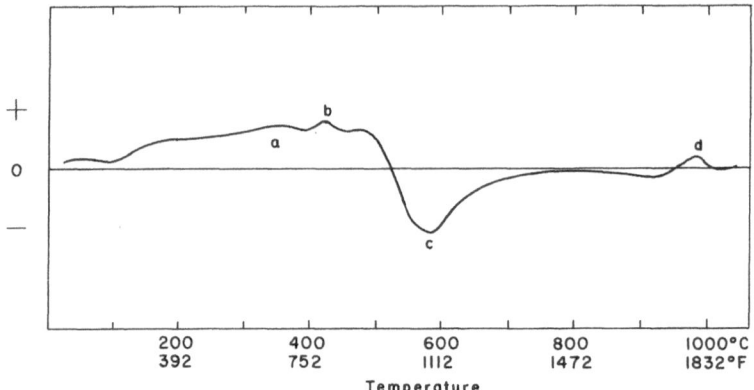

Fig. 74. Differential thermal analysis of a low-grade fireclay

The differential thermal analysis curve of Fig. 74 shows the reactions taking place on firing fireclays like the low-grade one described above. As the temperature is increased the curve shows carbon burnout at *a*, pyrite oxidation at *b*, dehydroxylation of clay minerals at *c*, and mullite nucleation at *d*. The mullite exothermic peak is greatly subdued over the one shown on Fig. 70, and this is caused by the lower degree of crystallinity of the fireclay type of kaolinite and the dilution by the other minerals present.

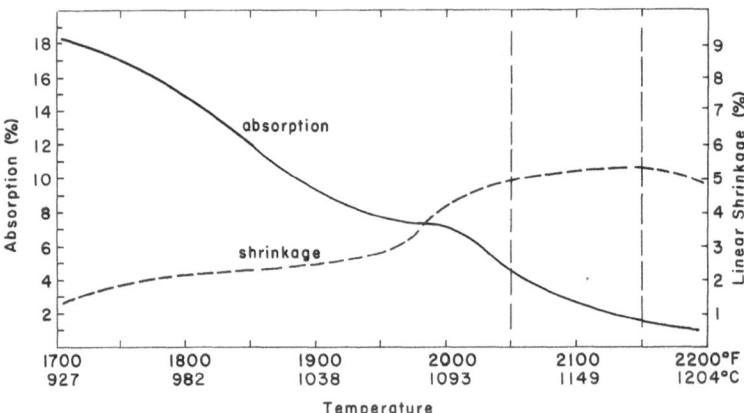

Fig. 75. Variation of fired properties produced by a fireclay typical of those used for structural clay products

On firing a clay body of this type, properties, such as water absorption on submersion and linear shrinkage, change as shown on Fig. 75. Open porosity as measured by absorption gradually decreases as firing temperature is increased while shrinkage increases due to consolidation, reactions, and melting. The decrease in shrinkage above 2150°F (1177°C) indicates overfiring through the formation of closed-pore bubbles in the body. Such behavior is called bloating and is usually noticed first by a reversal of shrinkage, even though distortion of the ware may not yet be visible. Soon after shrinkage reverses, absorption increases, bloating becomes visible, and the amount of liquid phase allows plastic flow of the ware under load. Of course, structural clay products cannot be fired into the temperature range where bloating occurs. The best raw material compositions produce rather flat shrinkage and absorption curves over a range of temperatures where the values of these properties are satisfactory and the desired color is obtained. For practical considerations it is well to have this temperature range as long as possible, since there is always a temperature differential in ceramic kilns, and it is necessary to have all the ware fired to acceptable properties regardless of temperature differences. To this end, the vertical dashed lines on Fig. 75 indicate the firing or maturing range for this particular fireclay. In commercial practice it could very well be that the temperature differential within the kiln will be much less than the 100°F range suggested here. Smaller differentials are all to the good as far as general uniformity of products is concerned.

The fired color of fireclays is buff, even with an iron oxide content as high as 5% Fe_2O_3. Of course this is a somewhat lower iron content than is usually encountered with red-firing, illitic clay bodies, but the quantity still does not explain the light buff color of fireclay products because, by judicious manipulation of the firing atmosphere, fireclay products can be made in pink and light red colors. The buff color can be explained on the basis of the ability of Fe^{3+} ions to enter the mullite crystals by substitutional solid solution. Fe^{3+} plays the role of Al^{3+} in mullite to a limited extent. When iron enters mullite, the white color of mullite is changed to light yellow. At 1100°C (2012°F) mullite can take 1.2% by weight of Fe_2O_3 into such a solid solution, 3.8% at 1200°C (2192°F), and 7.6% at 1300°C (2372°F). This solid-solution reaction between mullite and Fe_2O_3 removes the red hematite phase from the body when enough mullite has been developed to accommodate all the iron oxide present, thereby eliminating its characteristic red color [20].

Illitic clay and shale compositions can be expected to mature at lower temperatures than fireclays. Fig. 76 shows the changes in shrinkage and absorption with temperature for the hard, blue shale that was listed in Table 4A. The maximum firing range for good products including color is from 1900°F to 2000°F. Note that this material overfires at 2050°F (1121°C), and the characteristic increase in absorption after overfiring starts is visible. The rapid increase in shrinkage, and decrease in absorption followed by bloating above 2000°F, is characteristic of illitic clay bodies containing chlorite. The fired color of illite mixtures is always red on oxidation because they do not develop a sufficiently large mullite phase to accommodate all the ferric iron in solid solution, and since Fe_2O_3 is unreactive at these temperatures, it remains, giving its red color to the products. The phases present in this shale in the suggested firing range are mullite, quartz, hematite, and a trace of glass. Over the maturing range, mullite increases while quartz decreases, and the amount of glass increases slightly.

Some clay raw materials have carbonates of magnesium and calcium present in the form of magnesite, calcite, or dolomite, and these minerals tend to change the course of the high-temperature reactions to the extent that quite different products are produced.

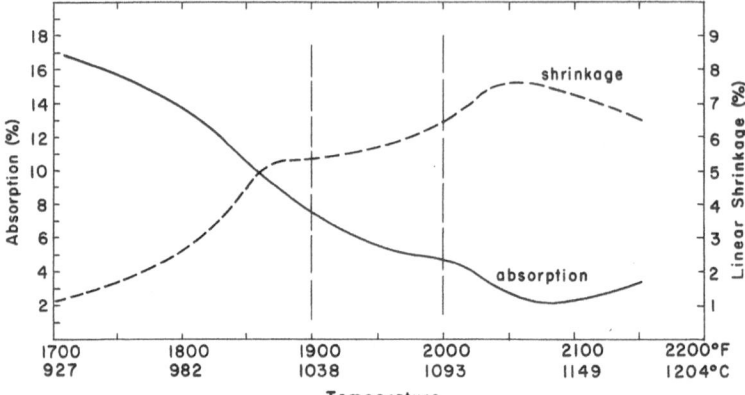

Fig. 76. Variation of fired properties produced by an illitic shale typical of those used for structural clay products

In spite of certain misgivings in the industry that have directed it away from the use of calcareous clays, these materials can produce unusually good products when treated properly. On heating up to 900°C (1652°F), the carbonates decompose with the evolution of carbon dioxide and leave very reactive oxides in the body. Solid-state reactions occur between the magnesium and calcium oxides and fine quartz particles and the aluminosilicates of the clays, which are, in themselves, very reactive at this temperature.

In solid-state reactions, two reactant particles must be in contact when enough energy in the form of heat is applied to make the reaction go. This is the activation energy. If a particle of CaO touches a particle of SiO_2 in this energetic state, the first product to be formed from the reaction is Ca_2SiO_4. As the reaction progresses, the product layer separates the reacting particles by an ever-increasing distance until one of the reactant particles is consumed. The reaction continues by the diffusion of cations through the reaction product layer, and since the diffusion path becomes longer with time, the reaction rate slows down exponentially. The process is expressed theoretically by the parabolic diffusion equation [21, 22]

$$\frac{dy}{dt} = \frac{k}{y} \tag{1}$$

where y is the thickness of the product layer, t is time, and k is a reaction-rate constant. The equation says that the rate of growth of the product layer is inversely proportional to its thickness. The integrated form of Eq. (1) becomes

$$y^2 = 2kt \tag{2}$$

Attempts have been made by several scientists to set up reaction-rate equations for powders by introducing various boundary parameters that seemed realistic, but none of them fit the situations found in clay products or, in fact, in most ceramic products [22, 23, 24]. The actual reaction rates in practical ceramics are always much slower at any given temperature than predicted by the rate equations developed so far. The reasons for this discrepancy are due to our inability to put quantitative terms on the probability of particle contact, grain-size distribution, particle shapes, and the extent and influence of surface diffusion.

The rate of high-temperature reactions is of great practical concern in the industry in order to assure completion of certain reactions in the least possible time. Probably the most important general principle of reaction rates to be appreciated by ceramists is the effect of temperature on the rate of all chemical reactions. The increase in rates with temperature is expressed by an exponential function which we call the Arrhenius equation. This equation is

$$\frac{d \ln k}{dt} = \frac{E_a}{RT^2} \tag{3}$$

where k is the reaction-rate constant as used in Eq. (1) and (2); T represents absolute temperature, E_a the activation energy, and R the gas constant. Integration of this equation gives

$$k = A \exp\left(-\frac{E_a}{RT}\right) \tag{4}$$

where A is the integration constant. From this one can see that the extent of reaction per mole in a unit of time is inversely proportional to the exponential value of the activation energy and directly proportional to the exponential of temperature; therefore, we must keep in mind that a slight increase in temperature causes a large increase in the rate of reaction.

The relations expressed in Eq. (2) and (4) become extremely important in the solid-state reactions involving CaO, MgO, and silicates. We shall also use the rate-temperature relation expressed in Eq. (4) a little later in the consideration of the proper temperature for oxidation reactions.

Calcium carbonate reacts with silica in a series of consecutive steps when excess silica is present, as is the case in structural clay products. The reaction starts as soon as the carbonate is decomposed, and certainly it will go to completion in time at 1000°C (1832°F) if the particle sizes of the reactants are small (around 20 mesh or 0.85 mm in dia.). There are three consecutive reactions in reaching the final stable phase, pseudowollastonite ($CaSiO_3$). In order of occurrence these are

$$2CaO + SiO_2 \rightarrow Ca_2SiO_4 \tag{5}$$

$$3Ca_2SiO_4 + SiO_2 \rightarrow 2Ca_3Si_2O_7 \tag{6}$$

$$Ca_3Si_2O_7 + SiO_2 \rightarrow 3CaSiO_3 \tag{7}$$

Eq. (5), the initial reaction regardless of the amount of CaO present, represents a fast reaction, but the reaction of Ca_2SiO_4 with silica is slow. The third step, presented as Eq. (7), is a fast reaction; therefore, the amount of $Ca_3Si_2O_7$ present at any time is small, and in properly constituted clay bodies its concentration is always negligible [25, 26].

This sequence of events is illustrated as Fig. 77 where the amounts of the various phases are given with respect to time at some constant temperature. Calcium oxide (a) disappears rapidly as a reactant while Ca_2SiO_4 (b) quickly becomes the major phase. Since the reaction to form $Ca_3Si_2O_7$ (c) is slow and its subsequent reaction with silica to form the metasilicate is fast, calcium pyrosilicate turns out to be a transient phase, coming and going without ever building up a substantial concentration. The final phase, $CaSiO_3$ (d), starts to form early in the sequence, and its concentration increases rapidly as the orthosilicate begins to disappear with time.

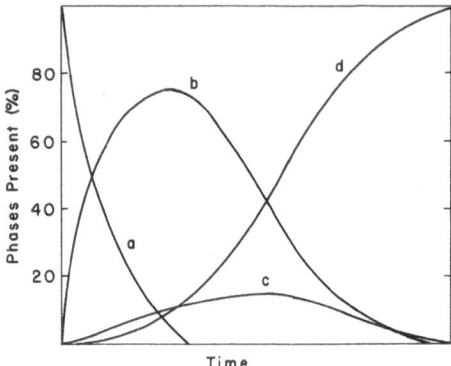

Fig. 77. Phases developed in the reactions between CaO and excess silica, curves for (a) CaO, (b) Ca_2SiO_4, (c) $Ca_3Si_2O_7$, and (d) $CaSiO_3$

Calcium oxide also reacts simultaneously with aluminosilicates in clay products. In addition to the above consecutive reactions with silica, gehlenite, $Ca_2Al(AlSiO_7)$, and anorthite, $Ca(Al_2Si_2O_8)$, are the products of lime reaction with dehydroxylated clay minerals. When the amount of CaO in the mixture is small (up to about 10%), anorthite is the phase produced as was previously described in Chapt. 5 with regard to the constitution of dryer scum; however, if the amount of CaO is 20% or more, gehlenite is produced as soon as the carbonate is decomposed. From the start of the reaction, the concentration of gehlenite increases to a maximum around 1100°C (2012°F). During this period gehlenite begins to react with silica from the clay and quartz at about 1000°C (1832°F) to form $CaSiO_3$, and the alumina from gehlenite is released to crystallize as corundum; consequently, a well-fired, high-lime body has quartz, gehlenite, pseudowollastonite, corundum, and hematite as final phases. Practically no fusions occur up to 1100°C; so there is no glassy phase in this type of product [27].

When MgO is a reactant in clay products as might be derived from magnesite, chlorite, or talc, the first reaction product is forsterite, Mg_2SiO_4, and it appears at about 900°C (1652°F). As heating proceeds, forsterite reacts with silica to form enstatite, $MgSiO_3$, directly. Structural clay products containing magnesia are not usually fired high enough to form the magnesium aluminosilicate, cordierite.

Dolomite is a double carbonate of calcium and magnesium, and when it is the source of CaO and MgO, diopside, $CaMg(Si_2O_6)$, is a reaction product; although, some anorthite may also be produced. These reactions take place on heating immediately after the release of the oxides from the carbonates, and free MgO is never found in clay bodies. This may be due to the fact that the double carbonate decomposes separately, and the MgO is released first to react with the dehydroxylated clay minerals.

In practical structural clay bodies such as those employed for structural clay products, the total amounts of calcium and magnesium oxides, as determined by chemical analysis, rarely exceeds 15%; however, the bodies can contain twice this amount before running into the danger of having excess free lime or magnesia in the product. The presence of these oxides in the fired product causes disasterous effects because both will hydrate on exposure to atmospheric moisture and convert back to the carbonate form by reaction with carbon dioxide in the air. This hydration and carbonation results in expansions of these phases within the product and complete disintegration will occur in a few days. For this reason the clay raw materials which happen to be interbedded with limestone or dolomite must be ground to 20 mesh or finer in order to prevent lime popping. In discussion of the rates of solid-state reactions, particle size was described as a controlling factor. Larger particles of the carbonates in the raw materials cannot completely react after decomposition because the product layer becomes too thick, and for all practical purposes the reaction stops before all the CaO has reacted. Such particles will soon cause localized expansions which fracture the body, and when they are located near the surface, spalling occurs and a circular section of the surface of the product breaks off. This is known as *lime popping*.

The fired color of high lime and magnesia products is buff, even though the basic clay mineral may be illite. From time to time the light colors have been attributed to the "bleaching action" of lime; however, this idea leads in an erroneous direction. CaO, MgO, or their silicates have no power to alter the red color of hematite. As a matter of fact, calcium and magnesium silicates are white in color and have no affinity for Fe_2O_3. As they become major phases in the product and the silicate crystals grow by continued reaction and sintering, they expel ferric ion oxide into grain boundaries. An examination of the microstructure will show points of hematite concentration where three silicate crystals join. This isolation of hematite into little pockets within the body causes the macroscopic visible appearance to be pink, buff, or yellow depending on the relative concentrations of the alkaline-earth silicates and hematite and on the extent of grain growth.

An example of the development of fired properties with an illitic clay containing dolomite and quartz is given in Fig. 78. This particular mixture contains 7.5% CaO and 6.6% MgO. The unusual characteristics shown are the maintenance of high absorption and low shrinkage until a temperature of $2100°F$ is reached, when melting occurs abruptly. Upon comparing the curves of Fig. 78 to those of Fig. 76, the action of the alkaline-earth oxides becomes strikingly apparent. The product appears to be more refractory at low temperatures and certainly more porous when calcium and magnesium silicates are present. Note also the narrower firing range for the production of satisfactory products.

The action of alkaline-earth oxides to reduce the amount of liquid phase produced below $2000°F$ can be used to good advantage in some cases. Illitic clays and shales that contain chlorite and sericite may develop a liquid phase at temperatures below the point where stable crystalline phases are produced. In such cases the shrinkage and absorption values are within acceptable limits, but the product will not withstand moisture and chemical attack as it should. Further firing of such products

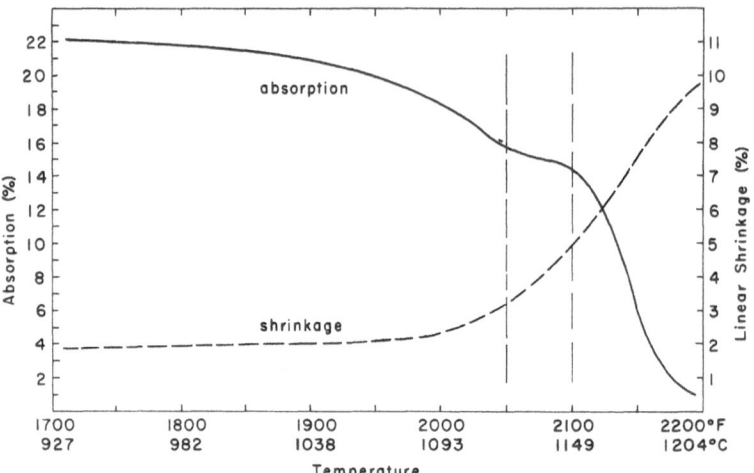

Fig. 78. Variations of fired properties produced by an illitic clay containing dolomite

produces too much liquid phase, and severe slumping and distortion occurs. The addition of just 3% CaO as $CaCO_3$ to this type of raw material extends the firing range at least 100°F. MgO or SrO can be used for the same purpose, and from the standpoint of quality, SrO is probably the best. The influence of 3% CaO to a low-temperature clay body is shown in Fig. 79. The curves in this figure were derived by heating a specimen under a load of 25 psi and measuring the change in height parallel to the applied load. Up to 1600°F, the expansion of the clay body without $CaCO_3$, represented by a dashed line, is due to thermal expansion, but the rapid subsidence

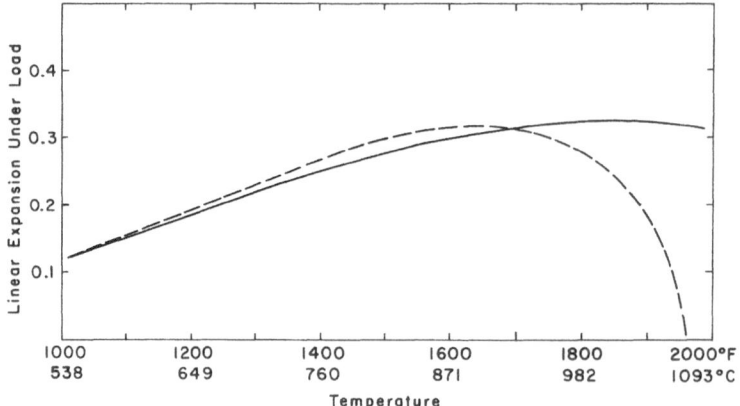

Fig. 79. The effect of a CaO addition to a clay body on the high-temperature slumping behavior. Solid line represents the addition and the dashed line without

thereafter is caused by a loss of strength as a liquid phase develops. This low-temperature melting and the resultant distortion under load was corrected by the CaO addition as is represented by the solid line. Now, the same clay products can be fired to 1900°F (1038°C) with equivalent absorptions and shrinkages without slumping under load. The higher temperature product will be more stable against various service conditions, and all of this is accomplished with a 3% addition of calcium oxide!

A caution is introduced here for those unaccustomed to dealing with calcium carbonate in the clay raw materials. This addition will immediately create dryer scum if the dryer is contaminated with sulfurous gases. Such a situation must be cleared up before pulverized limestone can be used as an additive to increase firing range and temperature. The substitution of strontium carbonate in this situation would alleviate this difficulty, but it may be more costly.

The reaction between calcium sulfate and the quartz-clay mixture should be of some concern to the structural clay products industry because it is related to the efflorescence problem and the contamination of dryer air with sulfurous gases. Calcium sulfate occurs in many clay and shale deposits in the form of gypsum, and it can be created in the body during the drying operation as was described in Chapt. 5. Incidentally, magnesium sulfate is of little concern in these matters, since it will completely decompose in one hour at a temperature as low as 1780°F (971°C); then MgO is free to react with the silicates as usual. On the other hand, calcium sulfate does not decompose by itself in the temperature ranges of most structural clay products. Calcium sulfate requires about two-and-a-half hours to completely decompose at 2400°F (1316°C), and the extent of decomposition in any reasonable length of time at 2200°F (1204°C) is negligible [28, 29].

Calcium sulfate will react with silica and aluminosilicates and liberate sulfurous gases at considerably lower temperatures than those required for simple decomposition [30]. An example of this type of reaction is expressed in the following equation:

$$2CaSO_4 + SiO_2 \rightarrow Ca_2SiO_4 + 2SO_3 \qquad (8)$$

Some particular cases of reactions between calcium sulfate and silica and an aluminosilicate are illustrated on Fig. 80. The solid lines show the extent of the reaction between calcium sulfate and cristobalite during the first hour at two temperatures of interest to us. It can be seen that the reaction is very slow at 1931°F (1055°C), but it goes nearly to completion in one hour at 2043°F (1117°C). The dashed lines indicate that the activation energy for the reaction of calcium sulfate with mullite is lower than that for cristobalite. The CaSO$_4$-mullite mixture reacts appreciably at 1872°F (1022°C) and goes practically to completion in 30 minutes at 2007°F (1097°C). These reactions were carried out on uncompacted powders with a large excess of cristobalite and mullite to increase the probability of reactant particles in surface contact [28, 31].

An examination of the results of the experiments described by Fig. 80, shows that there is a good possibility that some calcium sulfate will remain unreacted in some structural clay products where the maximum firing temperatures are from 1850°F (1010°C) to 1950°F (1066°C). These are typical firing temperatures for illitic clay products. If even a trace of the sulfate remains, it will cause efflorescence when the product is exposed to moisture due to the solubility of this salt. In addition, these reactions are going on with the evolution of sulfurous gases in the furnace zone of tunnel kilns; therefore, great care should be taken to prevent these gases from entering the waste-heat air for drying purposes. Since these reactions are slow around 1900°F (1038°C), there is a good possibility that they will continue to evolve sulfurous gases in the top of the cooling zone where the temperature is still high enough for the reactions. When this happens it would be impossible to keep sulfur contamination out of the dryer atmosphere if cooling air is used.

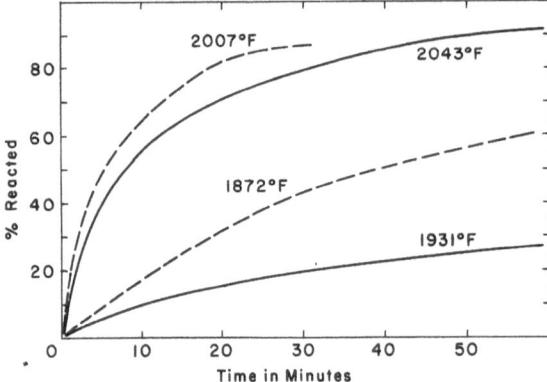

Fig. 80. $CaSO_4$-cristobalite and $CaSO_4$-mullite reactions at various temperatures. Solid lines are for the cristobalite reactions, and dashed lines represent reactions with mullite. After Brownell [28] and Jang [31]

When one learns how to handle clay raw materials containing alkaline-earth carbonates, excellent face bricks, structural tile, and quarry tile can be made. The basic principles can also be applied to carefully prepared floor and wall tile bodies; however, raw materials of this kind are not as appropriate for high-quality sewer pipes. The actions of the calcium, magnesium, and strontium silicates and aluminosilicates in clay products are to increase the coefficient of linear thermal expansion, reduce secondary moisture expansion to negligible values, maintain good resistance to the action of frost in spite of the increased absorption, and provide for narrower dimensional tolerances. The increased thermal expansion coefficient is of primary concern to those applying glazes to these products. It is a distinct advantage to have a body of higher thermal expansion if low-temperature glazes are to be applied. The reduction in long-range moisture expansion is due to the lower amount of glassy phase in these products.

6.3. Influences of Kiln Atmospheres

6.3.1. Kiln Atmospheres

Many solid-gas reactions are of great concern in the firing of structural clay products. The reactive gases from various sources constitute part of the kiln atmosphere. In order to obtain the product desired, these reactions must be controlled and this, in turn, means that the gaseous atmosphere in the kiln must be appreciated and regulated. In addition to the straightforward solid-gas reactions that occur within the products being heated, the gases in the atmosphere also affect the rates and paths of solid-solid and solid-liquid reactions.

More than any other factor, the importance of solid-gas reactions in firing clay products leads to the concept that kilns are chemical-reaction vessels and not simply heat-generating machines. This concept must be deeply ingrained in those responsible for quality products, since there is always a temptation to move straight to the most economical application of heat, which is not always compatible with the chemistry involved. In other words heat energy is not all that is required to change raw materials into useful ceramic products.

The gases of concern in firing structural clay products are CO_2, H_2O, CO, O_2, SO_2, and SO_3. All of these gases are introduced into the kiln atmosphere from the products of combustion of hydrocarbon fuels and by evolutions from the clay raw materials. We sometimes take for granted, but cannot ignore, the ambient air surrounding the whole process. For our purposes it is enough to consider air to be 23% by weight oxygen and 75% nitrogen, both of which are altered slightly by a variable amount of water vapor.

The massive introduction of combustion products into kilns makes this the primary source of CO_2, CO, H_2O, and O_2; therefore, the burning of fuel becomes the primary control over these gases. For example, the theoretical combustion of methane (CH_4), the principal ingredient in natural gas, is represented by the following equation:

$$CH_4 + 2O_2 \rightarrow CO_2 + 2H_2O \tag{9}$$

Since air is used as the source of oxygen, a great deal of nitrogen passes through as an inert gas. If more air is introduced with the gas than is necessary for complete combustion, residual oxygen will appear in the kiln atmosphere. When insufficient air is available to burn the hydrocarbon fuel, partial combustion can be achieved which results in carbon monoxide as a product. This situation is shown below.

$$CH_4 + O_2 \rightarrow CO + H_2O + H_2 \tag{10}$$

It is also possible to introduce so much fuel at a high temperature that some of the methane is simply broken down to free carbon and hydrogen, but such a situation should not be necessary in the firing of structural clay products.

The relations designated by Eq. (9) and (10) are illustrated on Fig. 81 [32]. Note that the zero point on the horizontal axis represents the exact amount of air for perfect combustion. The two combustion products at this point are CO_2 and H_2O. The dotted lines show the practical deviation from ideality. A mixture deficient in air creates carbon monoxide in the atmosphere, and an excess of air provides oxygen. It is of interest to observe that the carbon dioxide content goes through a maximum. Because the analysis of CO_2 is easily made, and as long as the fuel remains constant, it can be used as a monitor of the atmospheric conditions within the kiln. One needs to use caution in the interpretation of a CO_2 analysis, since a value of 9% could represent a deficiency or an excess of air; however, it is easy to determine which situation exists. By temporarily admitting a little more air, the CO_2 content will be seen to decline if excess air is present or the value will increase if a deficient air mixture prevails. These curves of Fig. 81 are for a particular natural gas; however, similar curves can be calculated for any hydrocarbon fuel using the above equations if an analysis is available for the particular fuel.

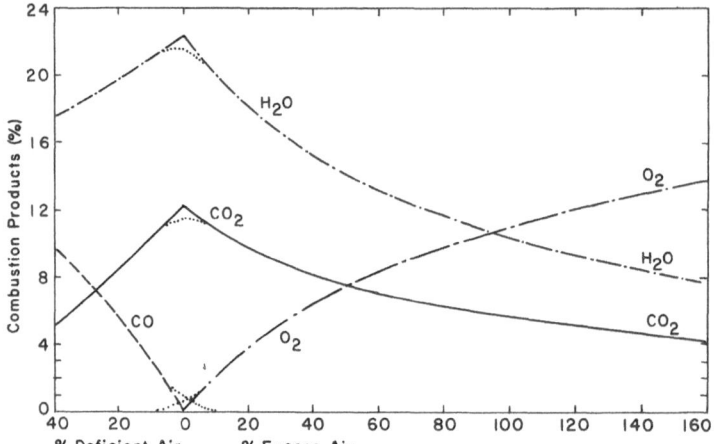

Fig. 81. Combustion products from a typical natural gas with respect to the amount of air available. North American Combustion Handbook [32]

Natural gas, as used in the United States, contains no sulfur compounds, but oils and coal most always contribute sufurous gases to the kiln atmosphere.

It has already been mentioned that clay raw materials also contribute gases to the firing environment. All clays evolve moisture as was discussed earlier in this chapter, and some materials give off CO_2, CO, SO_2, or SO_3 due to the oxidation of carbonaceous matter and pyrite or the decomposition of carbonates and sulfates. These evolutions must be considered for their possible effect on the atmosphere. During the dehydroxylation of clay minerals, about 14 pounds (6.35 kg) of water vapor is put into the atmosphere for every 100 pounds (45.4 kg) of clay. At the dehydroxylation temperature, this amount of water will occupy about 785 cu. ft. (22.2 m^3); therefore, it can act to reduce the oxygen content of the atmosphere, which may be undesirable as the ware goes from 850°F (454°C) to 1100°F (593°C).

6.3.2. Oxidation-Reduction

In the reactions involving gases with solids, both oxidation and reduction processes are utilized for different purposes and at different times in the firing operation. Since these chemical reactions are so important to product quality, the basic chemistry of oxidation-reduction reactions must be kept in mind. In the broadest sense, oxidation is a chemical change in which there is a loss of electrons. For example, if an atom of metallic iron with zero valence loses two electrons, it becomes a ferrous ion with a 2^+ charge, and the removal of the third electron creates a ferric ion with a 3^+ charge. With these valence changes the iron is said to be oxidized and is symbolically shown:

$$Fe^0 \rightarrow Fe^{2+} + 2e^- \rightarrow Fe^{3+} + 3e^- \tag{11}$$

Now, an atom or ion cannot simply discard its electrons without having a place for them to go, and in the oxidation process the electrons must be taken by some other element. Oxygen gas (O_2), for example, may take up the electrons lost by the iron; then, the oxygen becomes an ion with a 2^- charge. The process of gaining electrons is called reduction, and it can be expressed as,

$$2e^- + O^0 \rightarrow O^{2-} \tag{12}$$

The oxygen is reduced in the process of oxidizing the iron. In fact, any oxidation process is accompanied by a reduction; so it would be better to call these kinds of chemical reactions "oxidation-reduction reactions" in order that we do not lose sight of the chemical process.

In the firing of structural clay products, however, it is common practice to designate the reactions according to the chemical nature of the gaseous atmosphere. If elements in the clay body are being oxidized by the presence of oxygen in the atmosphere, it is called an *oxidizing fire*; and when carbon monoxide is part of the atmosphere, the firing is said to be *reducing* because of its effect on the elements in the ware. The important oxidizing reactions in the firing of structural clay products are represented by the following oxidation-reduction equations, in which the valences are indicated:

$$4Fe^{2+}O^{2-} + O_2^0 \rightarrow 2Fe_2^{3+}O_3^{2-} \tag{13}$$

$$4Mn^{2+}O^{2-} + O_2^0 \rightarrow 2Mn_2^{3+}O_3^{2-} \tag{14}$$

$$V_2^{3+}O_3^{2-} + O_2^0 \rightarrow V_2^{5+}O_5^{2-} \tag{15}$$

$$4Fe^{2+}S_2^{1-} + 15O_2^0 \rightarrow 2Fe_2^{3+}O_3^{2-} + 8S^{6+}O_3^{2-} \tag{16}$$

$$C^0 + H_2^{1+}O^{2-} \rightarrow C^{2+}O^{2-} + H_2^0 \tag{17}$$

Several reducing reactions of concern are

$$Fe_2^{3+}O_3^{2-} + C^{2+}O^{2-} \rightarrow 2Fe^{2+}O^{2-} + C^{4+}O_2^{2-} \tag{18}$$

$$Mn_2^{3+}O_3^{2-} + C^{2+}O^{2-} \rightarrow 2Mn^{2+}O^{2-} + C^{4+}O_2^{2-} \tag{19}$$

$$V_2^{5+}O_5^{2-} + 2C^{2+}O^{2-} \rightarrow V_2^{3+}O_3^{2-} + 2C^{4+}O_2^{2-} \tag{20}$$

In the above reducing equations it is the carbon which loses electrons; therefore, carbon is the reducing agent and is itself oxidized.

6.3.3. Oxidation of Carbon and Pyrite

Some clay raw materials contain carbonaceous matter that must be burned out as the temperature of firing is raised. If even a trace of carbon is present in the clay products as they reach the highest temperature, black coring and possibly bloating will occur as is illustrated in Fig. 82. This structural tile was cut in half with a diamond saw to expose the inside of the piece. On the outside it looked like a normal fireclay product except for the bulging, indicative of interior bloating. The amount of carbon left in this product is only a trace, located at a few points. The black color is due to the ferrous state of the iron present. Bloating is caused by gases being trapped in the vitreous body. It is possible to have black cores in the finished products without bloating, and this is still an undesirable condition. The FeO present during firing acts as a flux, and since it is not uniformly distributed, more glassy phase forms in one place than another within the product. Such a condition results in a wide range of strengths from one unit to another. In addition, the ferrous iron is more soluble in water and acid solutions than iron in the ferric state, which can cause a rusty iron stain on any building built from units with black cores. Finally, sulfides may remain within the reduced regions to create an efflorescence problem in service.

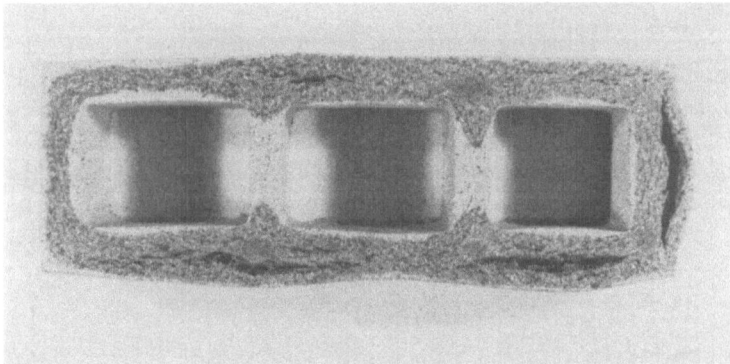

Fig. 82. Black coring and bloating visible in a fireclay product after sectioning (similar to that of Brownell [33])

It is possible to have a black core go through the high-temperature firing and be oxidized during cooling where the partial pressure of oxygen in the atmosphere is much higher. This can only happen with products which are relatively porous to allow for the accessibility of oxygen. If a black core is burned out during cooling, the process can be detected by examining the interior of the ware, because a core of a different red color will be visible. This is known in the industry as a *red heart*, and it is caused by the fact that the hematite (Fe_2O_3) phase experienced a lower temperature in the core than in the remainder of the body [33]. When the core is oxidized in the cooling period, there is a great chance that sulfurous gases will also be liberated to contaminate the products with adsorbed SO_3 molecules and to be carried over into the dryer with the waste-heat recovery. This situation occurs when pyrite is associated with the carbonaceous matter in the raw materials, and the fact that the sulfide cannot be completely oxidized as long as carbon is present. The sulfide is, then, also oxidized in the cooling period.

Since the removal of carbon is a time-consuming process, it is an economic advantage to accomplish the oxidation as quickly and expeditiously as possible. DTA, like that of Fig. 74, and TGA have indicated that carbonaceous matter in clays begins to oxidize around 570°F (300°C), which is previous to the loss of hydroxyl groups from the clay minerals [34]. After the crystalline water is removed, the whole clay body becomes very open and porous, which provides easy access for oxygen to penetrate the body and for CO and CO_2 to escape. Conditions within the body are, then, good for the oxidation process to proceed until oxide reactions start and shrinkage closes the open pore structure.

In order to determine the optimum temperature for carbon oxidation in clay products, one must recall that chemical reaction rates increase exponentially with temperature as designated by Eq. (4). The temperature at which an oxidizing atmosphere and time are provided for black core removal should be as high as possible to take advantage of the faster rates. As high as possible means a temperature just previous to the beginning of consolidation and shrinkage, after which gases cannot flow freely through the body. For red-firing products this temperature is about 1600°F (871°C), and for fireclay materials it is approximately 1800°F (982°C). During the oxidation period, it is also absolutely necessary to have 100% to 150% excess air when using hydrocarbon fuels—50% excess air is definitely insufficient and more air than prescribed does not help appreciably [33, 34].

The actual burnout rates for black cores in clay products are controlled by two antagonistic processes. Oxidation proceeds exponentially faster with temperature, but the diffusion of oxygen (into the body) and carbon gases (out) become exponentially slower with time because of the progressively longer diffusion path. Fig. 83 shows the exponential curve for the rate of core removal with time at a constant

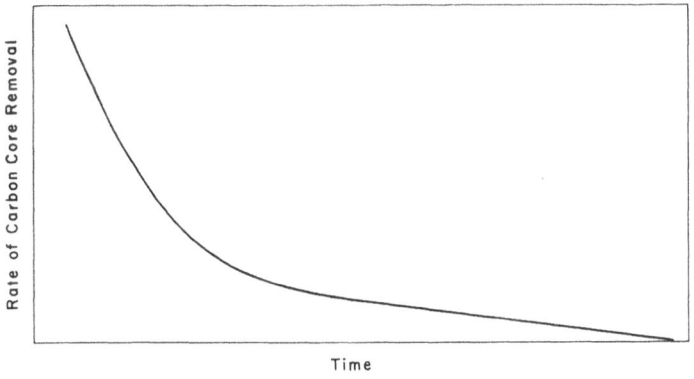

Fig. 83. Rate of oxidation of carbon cores from clay products, calculated from data by Brownell [33]

temperature, maximum body porosity, and 100% excess air. The data for this plot was taken from experimental results [33]. If one plots the time for complete black core removal against temperature, straight-line functions are obtained showing that the times are less for higher temperatures [35]. Such straight lines are the result of the two compensating exponential functions.

A few procedures have been proposed to assist the rate of carbon removal from clay products, but they cannot be presumed to take the place of any of the conditions already shown to be necessary. Water vapor in the kiln atmosphere at temperatures above 1300°F (704°C) reacts with carbon in the body according to Eq. (17), and it has been found that the introduction of extra water vapor during the oxidation period assists in the removal of the black core [36]. The addition of small amounts of ammonium chloride or hydroxide to the clay raw materials, previous to forming, expedites carbon oxidation. These additives appear to have a physical effect on core removal instead of chemical reactions. When these chemicals are added to the batch, the pore structure of the body is more open, and better diffusion paths for the gases are provided [33]. In a similar fashion, the rate of core removal is increased by additions of fillers, such as sand, grog, or calcined clay, to the raw materials. When the overall plasticity situation will allow appreciable additions of this kind, the resulting pore structure of the product is more suitable for diffusion, and the percentage of carbon is reduced at the same time.

Pyrite is a troublesome impurity found in many clay and shale deposits, especially those in which reducing conditions prevail. Dark gray, blue, and black clay deposits are most likely to contain pyrite, while red and yellow deposits almost never contain sulfides. The sulfurous gases emanating from pyrite during the firing process are primary sources of efflorescence on the finished product, and they present problems in environmental pollution.

Pyrite decomposes and oxidizes in the same temperature range as carbon, and oxidation is affected by the presence of carbon and water vapor. In the absence of oxygen the disulfide decomposes to the monosulfide at about 900°F (482°C) according to the following equation:

$$2FeS_2 \rightarrow 2FeS + S_2 \tag{21}$$

In the presence of oxygen, pyrite oxidizes in two steps which are indicated on DTA records in the form of two exothermic peaks. The first stage takes place at the decomposition temperature given above, and the second occurs at about 1090°F (588°C) [37, 38]. The equations for these sequential reactions are

(1) $$\qquad\qquad 2FeS_2 + 3O_2 \rightarrow 2FeS + 2SO_3 \tag{22}$$

(2) $$\qquad\qquad 4FeS + 9O_2 \rightarrow 2Fe_2O_3 + 4SO_3 \tag{23}$$

The overall reaction was given in Eq. (16). Evidences of the presence of pyrite in fireclay can be seen as small exothermic peaks superimposed on carbon-oxidation peak in Fig. 74.

In the presence of carbon, only the first step as given in Eq. (22) takes place. After the carbon has been oxidized, the second step in pyrite oxidation occurs. This is another reason why complete carbon burnout should be accomplished before high temperatures are reached. If a carbon core is taken up to the maximum firing temperature and into the cooling period, there will be sulfurous gas emissions at these times.

Water vapor assists in the decomposition of pyrite in much the same manner as it did in carbon removal. The following equation expresses this reaction:

$$3FeS_2 + 2H_2O \rightarrow 3FeS + 2H_2S + SO_2 \tag{24}$$

As in the water-carbon reaction, complete oxidation can only be achieved with the presence of oxygen, but the presence of water vapor appears to speed up the decomposition process [38].

Since sulfurous gases are evolved into the kiln atmosphere at low temperatures by the oxidation of pyrite, at high temperatures by the reactions of sulfates, and throughout the firing if a sulfur-containing fuel is used, the SO_2-SO_3 equilibrium is an important relation to be understood. Both oxides react with alkali silicates to form sulfites and sulfates, and SO_3 has a greater tendency to be adsorbed on silicate surfaces than SO_2. The reaction expressed by Eq. (25) is exothermic; therefore, the equilibrium is favored in the direction of SO_2 when heat is applied.

$$2SO_2 + O_2 \rightleftharpoons 2SO_3 + heat \tag{25}$$

The equilibrium volume percentages of these two sulfurous gases, when the partial pressure of oxygen remains constant, are given for temperatures of interest to the structural clay products industry by Fig. 84 [39, 40]. One can readily see that the sulfurous gas

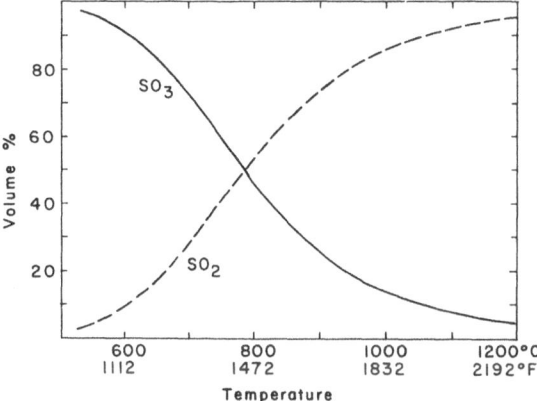

Fig. 84. Volume ratios of SO_3 and SO_2 across a significant temperature range assuming a constant partial pressure of oxygen. Calculated from data by Lovejoy [39] and Mellor [40]

produced by the oxidation of pyrite will be mostly SO_3, and that produced from the reaction of sulfates at high temperatures will be largely SO_2. It also shows that if sulfurous gases are being evolved by black-core oxidation in the cooling period at, say, 1500°F (816°C), there will be nearly equal amounts of both oxides in the atmosphere.

6.3.4. Color Development and Control

Iron oxide is the most important and flexible colorant in all structural clay products. It is present in most all clay raw materials from 1% to 8%. In some cases extra iron oxide may be added to intensify its coloring effect. When reduced by carbon monoxide, the form of iron oxide can be either magnetite (Fe_3O_4) or wustite (FeO) both of which are black at all temperatures. In the highest oxidation state, hematite (Fe_2O_3) is produced, and its color is temperature dependent. Hematite derives its color from a defect crystal structure, and the number of defects per unit of volume increases with temperature. The development of defects is a reversible reaction, but the rate of removing defects is very much slower in air than the production of defects on heating; therefore, it is easy to cool a product locking in the defect structure obtained at the highest temperature. At very low temperatures Fe_2O_3 is almost orange in color, and as the temperature is increased darker reds are produced until the color is almost black at 2400°F (1316°C). This means that the development and control of color is extremely dependent on the highest temperature the product reaches.

Reduction of iron oxide at the highest temperature is often used to produce tan, brown, and black products from red-firing materials. The procedure is called *flashing*, and it is accomplished by reducing the air supply to the deficiency side of perfect combustion by either decreasing the air supply to the fuel or introducing excess fuel. After flashing, the ware goes immediately into the cooling period where atmospheric air provides a much higher oxygen content, and there is a tendency to reoxidize the exposed surfaces back to shades of red. If black products are desired there are a few tricks that can be employed to "hold the flash" while cooling. The reduced state can be maintained if the top firing temperature is high enough to flux the FeO into a glassy phase. Of course these products will have very low absorption. The best blacks are produced by adjusting the firing atmosphere close to theoretical combustion as soon as the oxidation period is passed, then flashing for 10 minutes at the finish temperature. Another procedure to hold the black color produced by flashing at lower maturing temperatures is to introduce a 5% methanol solution into the kiln during the entire cooling period until the temperature is reduced to 850°F (454°C) [36]. Sometimes flashing is done at the top temperature with no intention of maintaining the black color. The reoxidation during cooling produces colors that are more tan and brown than the typical brick red obtained by good oxidation.

When one thoroughly understands the behavior of iron oxide in firing clay products many kinds of manipulations can be employed to produce special effects. For example, a pink fireclay product can be made. It was already mentioned that the mullite phase of such products takes ferric iron into solid solution at high temperatures, and this accounts for the usual buff shades. By firing fireclay products under a slightly reducing atmosphere and maintaining the environment while cooling to 1800°F (982°C), the iron oxide is prevented from going into solid solution with mullite. A highly oxidizing atmosphere is introduced at 1800°F and the temperature held for a few hours in order to oxidize the iron oxide to the ferric state. The iron oxide will produce a light red color characteristic of its 1800°F temperature. Other manipulations of the valence state of iron oxide in clay products produce special effects.

Body stains of various kinds are sometimes used for various colors and shades, and some of these are sensitive to atmospheric conditions. Manganese compounds are frequently added to produce gray fireclay products and black to brown illitic clay products. When the mullite phase in a product is low and the iron oxide content high, the effect of manganese is to produce brown colors.

6.4. Types of Kilns

Continuous tunnel kilns are most commonly used throughout the structural clay products industry because of their contribution to reduced costs of production;

however, periodic kilns are still used in special cases. Some old-style periodic kilns have been modernized and new designs have been developed for the industry. Old-style periodic kilns with old-style burner and draft systems have no place in the industry today.

Tunnel kilns are complicated pieces of machinery, and they require competent and alert ceramic engineers to keep them operating at full capacity without degeneration of their various systems. Size alone is indicative of the magnitude of engineering knowledge required for their design and operation. Tunnel kilns in the structural clay products industry range in length from 300 ft. to 650 ft. (91 to 198 m) long, and each one is custom-built to fit the demands of the raw materials and economic production rates. The complex draft situation within the kiln, as described in Chapt. 5.4.2, must be monitored and controlled, and the basic movements of gases is further complicated by recirculation circuits designed to reduce temperature differentials within the load. Complex burner systems provide any kind of atmosphere desired in the various zones of the kiln when their operation is thoroughly understood. Fig. 85 shows part of a tunnel kiln designed for the production of face bricks. The high-temperature furnace zone is visible to the left and the cooling zone to the right. The large air ducts overhead are for removing heated fresh air from the cooling zone and delivering it to the dryer.

Fig. 85. Tunnel kiln for the production of face bricks. Courtesy of Swindell-Dressler Co.

Tunnel kilns are designed to fire ware rapidly on a particular inflexible schedule. This means that optimum results are obtained when the raw materials and product size remain constant. Color variations, as often required in the manufacture of face bricks, should be accomplished with various types of coatings. Although continuous kilns are not especially suited to frequent schedule changes, modifications are made from time to time in order to accommodate different products. Whenever a change is introduced, several hours are required for stabilization on the new conditions.

Tunnel kilns make the best possible utilization of heat if they are well insulated against conduction heat losses. The heat exchange is so good that the exhaust gas temperatures are down to 300°F (149°C) to 400°F (204°C). This, of course, makes these kilns efficient users of fuel; however, good maintenance is required for the prevention of leaks in the draft system. All-in-all a well-maintained kiln will operate continuously for years without requiring complete shutdown for repairs. Part of this longevity is due to holding the temperature constant so that stresses of thermal expansion are eliminated.

The only parts of the tunnel kiln complex that experience thermal fluctuations are kiln-car decks; consequently, they require special materials and constant care. Routine maintenance of car decks on every cycle through the kiln has become a necessary practice with the introduction of automatic setting equipment. The characteristics of the refractories for car decks are high-temperature strength, moderate impact resistance, low thermal expansion, and moderate resistance to thermal shock [41]. The design of car tops varies widely even within the two basic types. Fig. 86 gives a good view of a car deck on an especially wide car emerging from the kiln

Fig. 86. Kiln car emerging from a wide tunnel kiln. Courtesy of Swindell-Dressler Co.

illustrated in Fig. 85. The metal plate for completing a convection seal and the radiation shield of overhanging refractories are clearly visible along the lower side of the car. Note that the brick are stacked so that the burner flames will pass across the car in a corridor above the car deck. This arrangement prevents flame impingement on the ware. Sometimes tunnel kilns are designed to fire under the top deck, and a car of this type can be seen in Fig. 55. In this case, the refractories must withstand direct impingement of flames and very high temperatures.

There are three basic types of periodic kilns in use in the structural clay products industry. They are large rectangular updraft and downdraft kilns, round downdraft (beehive), and shuttle kilns that are essentially updraft in operation. All of these kilns should be equipped with forced-air burner systems where the air-gas ratio can be easily adjusted to provide periods of excess air, no excess air, and deficient air over a complete firing cycle. When so equipped, the kilns can be operated on a positive pressure and no stack fan for forced draft is required [42]. The fan on the burner system can be used to control the stack draft during firing and cooling. Dampers on the exhaust system are used to assist in maintaining the desired atmosphere through draft control. The capacity of a periodic kiln is about equivalent to one day's production from a tunnel kiln.

The round downdraft kiln shown in Fig. 87 is used for firing large sewer pipes and fittings. The floors in downdraft kilns are constructed as a refractory grating, opening into radial flues connected to a central flue that leads to the exhaust stack. If waste-heat recovery is intended, a duct is connected to the main flue before it reaches the stack, and a set of dampers is used to control which of the two paths the heated air takes. Dried ware is set in round and rectangular kilns by means of a fork-lift truck. The heavy pipes are handled within the kiln by means of a temporarily constructed electric hoist.

Fig. 87. Round downdraft kiln used in the sewer pipe industry

Shuttle kilns are like the furnace section of a tunnel kiln, and the ware is stacked on kiln cars similar to those used in continuous firing. A full load of dried ware is placed in the kiln; the doors are closed, and the firing temperature is controlled with respect to time. A large shuttle kiln is shown in Fig. 88. It is equipped with a forced-air burner system in which the gas-air ratio is controlled.

Periodic kilns find a place in the structural clay products industry where specialty items are a part of a company's market line. These are usually small orders requiring variations in raw materials and firing schedules. Odd sizes and shapes fall into this category also, and they can be handled best in a periodic firing situation. The two-fire glazing process is often accomplished in shuttle kilns unless the demand can keep a small tunnel kiln in production. There is no doubt that the cost of firing is higher with periodic kilns than with the tunnel kilns, but this is compensated for by the higher price obtainable for specialty items.

Fig. 88. Shuttle kiln with a positive pressure burner system. Courtesy of Swindell-Dressler Co.

A look at the characteristics of periodic-kiln firing discloses some advantages as well as disadvantages. The turnover time for these kilns in this industry is from 2.5 to 6 days depending on the kiln and the products being fired. The large round down-draft kilns have a temperature distribution problem, and much of the firing time on these kilns is spent in trying to equalize the temperature from top to bottom—especially the bottom center. On the other hand, shuttle kilns give the closest control over temperature, time, and atmosphere of all kilns including tunnel kilns. Cooling schedules can also be precisely controlled. Because periodic kilns of all types are cycled through large temperature ranges, the repair and maintenance of refractories is more extensive than in tunnel kilns. All-in-all, the best shuttle kilns provide precision firing at an extra cost.

6.5. Kiln Firing

A typical firing curve for structural clay products is shown on Fig. 89. The various rate periods have been plotted proportional to time or kiln length, since the charging rate of a tunnel kiln determines the firing time. The actual times in the periods will vary from one material to another, but the curve presented here could be for a rather high-temperature, red-firing material or a low-temperature fireclay. The oxidation hold around 1550°F (843°C) could be about 6 hours, or in some cases it may require 12 to 24 hours in this period. In the beginning, the temperature is brought up to the oxidation at about 200°F per hour. Not all materials require an oxidation delay, especially if plenty of excess air is used across the temperature range involved; consequently this part of the firing curve would be deleted and the final temperature obtained sooner. After the oxidation has been completed the temperature should be raised as rapidly as possible, without too much differential within the setting, to the maximum. A hold of at least 2 hours should be made at or close to the top temperature to allow for reaction rates to slow down. Ordinarily, not much phase development occurs after a 2-hour period except when high-lime bodies are being fired. Hold periods of from 8 to 48 hours are required for the desirable lime-silicate reactions to be completed, and the exact time of hold would depend on the top temperature, with shorter times at higher temperatures. Cooling can be carried out as rapidly as possible without causing damaging cracks.

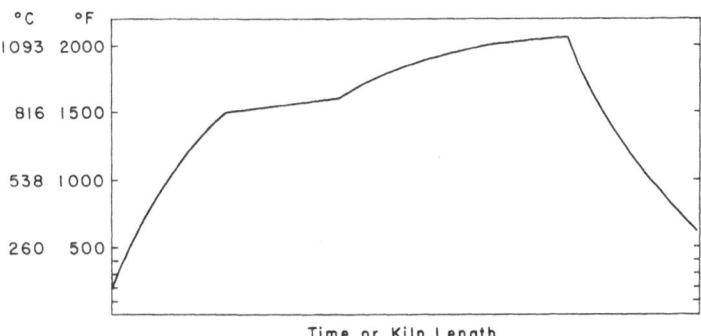

Fig. 89. Typical firing schedule for structural clay products

Environmental air pollution problems are being encountered throughout industry, and they are likely to remain. It is unnecessary for ceramic kilns to exhaust particulate matter into the atmosphere, but sulfurous gases are unavoidable.

Flashing has been the main cause of particulate pollution for many years, and this has resulted from a lack of knowledge as well as controls. Reduction or flashing in tunnel kilns has been less of a problem, since the incomplete combustion products are usually consumed while the furnace gases are preheating the incoming ware. With periodic kilns, however, black smoke containing carbon particles is often expelled from exhaust stacks during periods of reduction. Such emissions are unnecessary, since it is carbon monoxide gas that is the reducing agent, and this is an odorless colorless gas. When devices are not used to measure, directly or indirectly, the CO content of the kiln atmosphere, a reduction atmosphere is guaranteed by the visible appearance of free carbon in the exhaust; however, maximum CO concentrations can be obtained by introducing enough air to oxidize the free carbon to carbon monoxide. When carbon particles are a part of the exhaust, they can be oxidized in the flue by an after-burner before they are sent up the stack.

At the present time emissions of sulfur dioxide and trioxide can only be controlled by the use of scrubbers. They pass the exhaust gases through an alkaline spray to collect SO_3, and react SO_2 with pulverized lime or limestone. In the process, SO_3 is converted to calcium sulfate, and SO_2 is combined to form calcium bisulfite, both of which are removed as solids [43]. Sulfur compounds are so common in clay raw materials and fuels that it appears to be an unavoidable pollution problem. For reasons already set forth, it is undesirable to retain sulfur compounds in the ceramic products; so they must be removed from the gaseous exhaust.

6.6. Burner Systems

The importance of atmospheric control in the firing of kilns has been emphasized, and there are several burner systems available to achieve this end. Of course simple inspirating burners can be used to fire kilns when fans are used on the exhaust stack to provide a means of controlling draft, but the more complex systems with fuel-air control are so far superior that little attention will be given here to the former. With the old inspirating burners, a kiln could not be fired oxidizing under a positive pressure, but the newer systems to be described here have no problem in this regard. In addition flame temperatures can be throttled down to around 1200°F (649°C) through the use of excess air and good mixing.

There are basically two types of gas-air burner systems in use, but there are several modifications on each. One system uses a controlled gas-air ratio burner, and the other premixes the gas and air before delivery to the burners. Combustion mixtures varying from deficient to excess air can be obtained with both systems. Burner modifications are possible to provide a variation in flame velocity and length, from a flat radiant flame to high-velocity jet flames for great penetration into the center of kilns. There is little or no flame impingement with flat-flame burners, but jet burners must have open channels to prevent impingement of the flame on the ware. High-velocity jet burners should not be used where dusting is a problem. With structural clay products kilns, they would probably not be suitable for glazed ware. The principal advantage of jet burners is to provide more even heat distribution throughout a large kiln.

Two appropriate modifications of the controlled gas-air ratio burner system, usually called excess-air burners, are shown in Figs. 90 and 91. Both drawings show the gas coming in through a shut-off cock, pressure regulator, and pressure switch for monitoring the pressure on the system, the air supply being provided by a blower or fan. All of the pressure switches are wired to close the manual-reset safety switch in case of an improper situation. The dashed lines on the drawings represent a safety feature that prevents turning on the gas supply while any burner is open. This is accomplished through supervisory gas cocks in the line to each burner. Several burners can be operated in parallel after the gas-air regulator, but only one pressure switch is required at the end of the air line through the supervisory gas cocks. The system designated in Fig. 90 provides a constant air supply to the burners, and the amount of gas fed to the burners is controlled by the demand of the temperature controller (T.C.). This system provides a variable amount of excess air depending on how hot a flame is required, but a range of excess air can be provided by adjustments to the bleeder and the gas-air regulator. Such systems are used on periodic kilns as well as

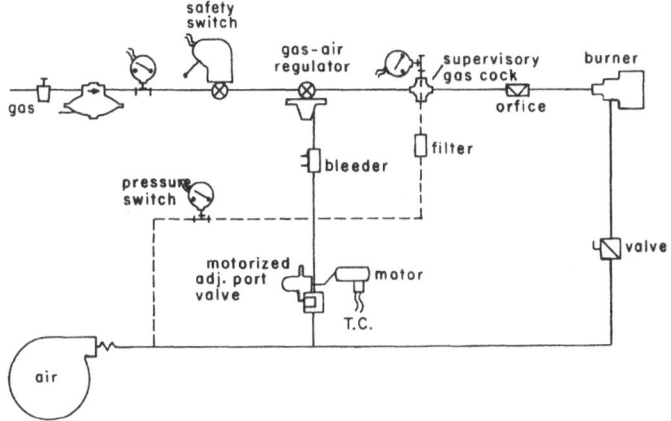

Fig. 90. Excess air burner system with constant air supply to burners

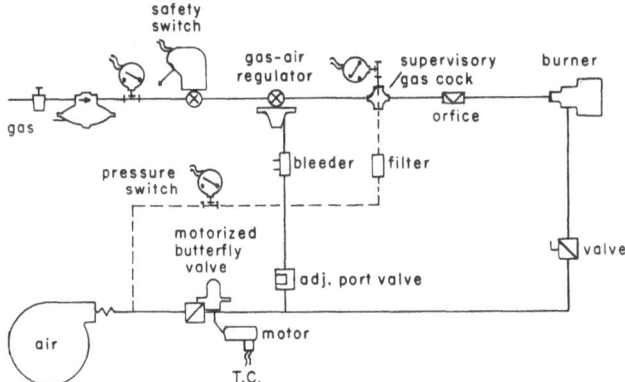

Fig. 91. Excess air burner system with constant gas-air ratio at burners

on the furnace zones of tunnel kilns. When the amount of excess air should remain constant while controlling temperature, a system such as that in Fig. 91 is used. Here the air to the burners is controlled, and the correct amount of gas is mixed by adjustments to the port valve, bleeder, and gas-air regulator. In this case the size and velocity of the flame varies in response to temperature-controller demands. A constant excess-air system might be used in the oxidation, preheat zone of a tunnel kiln. Both systems can be adjusted for a deficiency of air as well as excess air. Excess air burner systems can be seen as installed on Figs. 85, 87, and 88.

A typical premix burner system is sketched on Fig. 92. Atmospheric air is sucked in through a filter by a compressor, and gas is brought in through high-pressure, low-pressure, and manual-reset safety switches. The gas governor delivers gas to the mixer at the same pressure as the air. By means of a bypass system, the gas-air velocity

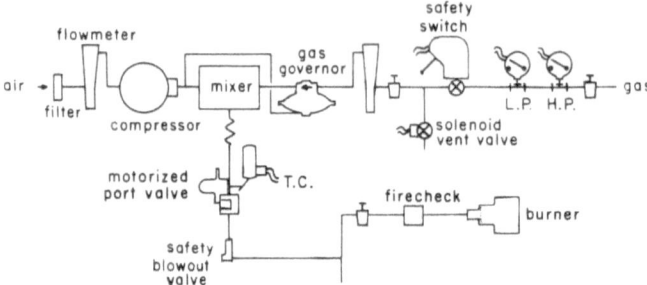

Fig. 92. Premix burner system for gas and air. Similar to Selas Corporation of America

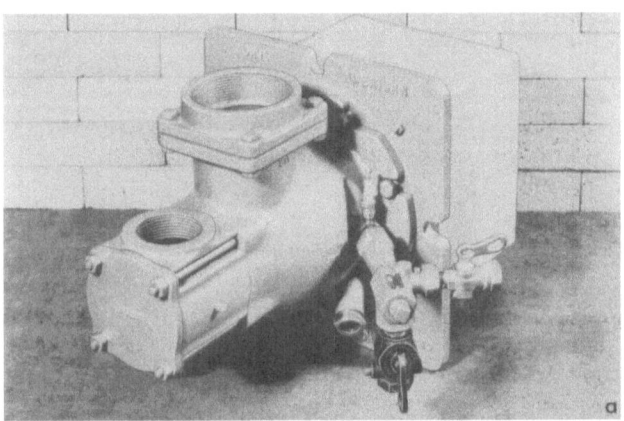

Fig. 93. (a) Gas-air excess air burner equipped with pilot and spark ignition.
(b) Dual-fuel burner with oil valve attached. Courtesy of The North American Manufacturing Company

is maintained constant through the mixing valve to insure adequate mixing. The gas-air ratio is set at the mixer by the aid of flowmeters, and the amount of premixed fuel delivered to the burners is controlled by a motorized, adjustable-port valve. In this system a safety blowout valve and firechecks to stop flashbacks must be installed in the premix line. A firecheck is required to be installed near each burner. This type of system is ideal for periodic kilns or the furnace zones of tunnel kilns where a fixed air-gas ratio can be maintained. The premixing provides for fuel economy through proper combustion.

Fig. 94. Dual-fuel burner system installed on a top-fired round downdraft kiln. Courtesy of The North American Manufacturing Company

Gas is the most convenient fuel to fire structural clay products, but its limited supply makes the use of other hydrocarbon fuels necessary. To meet this problem, dual-fuel burner systems are available in the excess-air type. A separate oil line is brought through the system to each burner and valves installed to quickly convert from one type of fuel to another. Close-up photographs of an excess-air burner and a dual-fuel burner are shown in Figs. 93 (a) and (b), respectively. Pilot burners equipped with spark ignition are attached to both burners, and the refractory blocks that are inserted into the kiln wall are shown to the right-hand side. The dual-fuel burner is equipped with an oil valve extending to the left. Of course, a gas valve will be installed near each burner in the gas line. A practical installation of a dual-fuel burner system is given in Fig. 94. The system is installed on a top-fired round downdraft kiln. Several elements, symbolized on the line drawings of Figs. 90, 91, and 92, may be seen here.

6.7. Cooling Stresses

The ultimate strength and uniformity of strength values for structural clay products is dependent on the phases present and the cooling rate. The phases each contribute their own thermal expansion to the overall expansion of the product in proportion to the amount present. It is the thermal expansion, or rather contraction, which on cooling sets up stresses in the product which can lead to loss of strength through cracking. The faster cooling rates promote greater temperature differentials within the products which, in turn, leads to greater stresses. The magnitude of the stress development at any particular temperature differential is determined by the thermal expansion.

Now, structural clay products are far from being homogeneous bodies. They may have from 4 to 8 individual phases of primary importance to final properties, and it may very well be that no two factories produce exactly the same product. Just because the constitution of these clay products is complex is no reason why there should be little concern about the phase composition. It has been found that when two phases in a rather dense body have differences in the coefficients of linear thermal expansion of 40×10^{-7} cm/cm/°C or greater, the weaker phase will be fractured by cooling stresses. Since the particle sizes of the phases are small, the fractures are called microcracks. Two phases which were shown to be compatible had a difference in their coefficients of 4×10^{-7} cm/cm/°C [44]. From this, one can begin to appreciate the importance of putting phases together which do not tear the microstructure apart.

Microcracking is the start of complete product failure. When microcracking occurs, the strength of the ware is lowered, and the strength values become erratic. Extensive microcracking occasionally adds up to macrocracking when the whole piece is weakened by fortuitously located microcracks. In any case, a product so weakened is said to have been dunted, and the situation is noticed by striking two articles together. A dunted product gives a dull thud instead of the ringing sound characteristic of good strength.

Table 13. *Average Coefficients of Linear Thermal Expansion for Phases Commonly Found in Structural Clay Products*

Phase Name	Formula	Coefficient (cm/cm/°C)
Corundum	$a\text{-}Al_2O_3$	88×10^{-7}
Quartz	SiO_2	120
Mullite	$Al_4O(Al_2Si_2O_{12})$	53
Pseudowollastonite	$CaSiO_3$	96
Anorthite	$Ca(Al_2Si_2O_8)$	43
Forsterite	Mg_2SiO_4	98
Clinoenstatite	$MgSiO_3$	82
Diopside	$CaMg(Si_2O_6)$	68
Cordierite	$Mg_2Al_3(AlSi_5O_{18})$	11
Albite (feldspar)	$Na(AlSi_3O_8)$	90
Hematite	Fe_2O_3	94
Magnetite	Fe_3O_4	87
Glass	$Na_2O\text{-}CaO\text{-}SiO_2$	90
Fused silica	SiO_2	5

In order to predict phase compatibility as far as thermal expansion is concerned, Table 13 is presented. The mean coefficients of linear thermal expansions are listed for several phases of interest to those working in the structural clay products industry. From these data it is not difficult to see that the common mullite-quartz-glass body is beset with microcracking difficulties if one tries to cool it rapidly. On the other hand, a phase assemblage of pseudowollastonite, diopside, corundum, and hematite, such as results from the use of dolomitic clays, could be cooled much more rapidly without extensive damage.

Considering the realities of structural clay products with regard to phases, particle sizes, and porosities, the general shape of the cooling curve to avoid catastrophic failure can be determined. The resistance to failure of the overall body can be expressed as

$$R = \frac{Ma}{Ee} \tag{26}$$

where M is the modulus of rupture, a the thermal diffusivity, E the modulus of elasticity, and e the coefficient of linear thermal expansion. Using this relation the maximum rate of cooling without complete thermal failure is

$$\frac{dT}{dt} = -K\frac{Ma}{Ee} \tag{27}$$

where T is temperature; t represents time, and K is a size-shape factor, constant for a given system.

The integral form of Eq. (27) may be written as

$$t_2 - t_1 = -\frac{1}{K}\int_{t_1}^{t_2}\left(\frac{Ee}{Ma}\right)dT \tag{28}$$

Since Ee/Ma are all functions of temperature, and the particular equations for these relations are not available, they are determined experimentally; then the integral can be graphically evaluated by plotting Ee/Ma as a function of temperature. From this type of solution, a safe relative cooling schedule can be plotted. The actual times are then found for a particular product by trial-and-error while conforming to the general shape of the calculated cooling curve [45].

Calculations such as these have been carried out on several structural clay products, and they show that the cooling rate should be slower after the glassy phase has become rigid and during the quartz inversion [45]. For the most part, the practical cooling rates of kilns in commercial use are slow enough not to require a special shape; however, if faster cooling rates are desired, these calculations will help to determine a safe schedule.

The new thermal-acoustical analyzer appears to be another practical approach to the determination of safe cooling schedules for particular products. With this apparatus one can determine the temperatures of sensibility and the temperature differentials when microcracks and macrocracks occur. From this direct information, cooling rates can be established for various periods during the cooling process [46].

It appears that microcracking is to be a phenomenon in structural clay products for the foreseeable future, since properly made products have ample strength for the purposes intended. Attention should be given to controlling the microcrack structure in order to prevent complete failure of the products. Hasselman [47] has given much attention to crack propagation through ceramic bodies. He has found that weaker materials with relatively high densities of long microcracks can be subjected to considerably more severe temperature differentials before complete failure occurs than can stronger materials with short microcracks. This describes fairly well the state of affairs in structural clay products where long microcracks, a wide particle-size distribution, and pores all act to limit crack propagation.

References

1. Brindley, G. W., and M. Nakahira: Kinetics of dehydroxylation of kaolinite and halloysite. J. Am. Ceram. Soc. **40**, 346–50 (1957).
2. Brindley, G. W., and M. Nakahira: The kaolinite-mullite reaction series: II, Metakaolin. J. Am. Ceram. Soc. **42**, 314–18 (1959).
3. Brindley, G. W., and H. A. McKinstry: The kaolinite-mullite reaction series: IV, The coordination of aluminum. J. Am. Ceram. Soc. **44**, 506–7 (1961).
4. MacKenzie, K. J. D.: Infrared kinetic study of high-temperature reactions of synthetic kaolinite. J. Am. Ceram. Soc. **52**, 635–37 (1969).
5. Roy, R., D. M. Roy, and E. E. Francis: New data on thermal decomposition of kaolinite and halloysite. J. Am. Ceram. Soc. **38**, 198–205 (1955).
6. Brindley, G. W., and M. Nakahira: The kaolinite-mullite reaction series: III, The high-temperature phases. J. Am. Ceram. Soc. **42**, 319–24 (1959).
7. MacKenzie, K. J. D.: Infrared frequency calculations for ideal mullite ($3Al_2O_3 \cdot 2SiO_2$). J. Am. Ceram. Soc. **55**, 68–71 (1972).
8. Comeforo, J. E., R. B. Fischer, and W. F. Bradley: Mullitization of kaolinite. J. Am. Ceram. Soc. **31**, 254–59 (1948).
9. Comer, J. J.: Electron microscope studies of mullite development in fired kaolinites. J. Am. Ceram. Soc. **43**, 379–84 (1960).
10. Glass, H. D.: High-temperature phases from kaolinite and halloysite. Am. Min. **39**, 193–207 (1954).
11. Verduch, A. G.: Kinetics of cristobalite formation from silicic acid. J. Am. Ceram. Soc. **41**, 427–32 (1958).
12. Phillips, G. C., Jr.: Behavior of Illite on Heating, M.S. Thesis, New York State College of Ceramics, Alfred University, June 1964.
13. Tatem, W. A.: The Melting of Illite, M.S. Thesis, New York State College of Ceramics, Alfred University, June 1960.
14. Brindley, G. W., ed.: X-Ray Identification and Crystal Structures of Clay Minerals. London: The Mineralogical Society. 1951.
15. Kellogg, Alan E.: Changes in Prochlorite on Heating to 1000°C, B.S. Thesis, New York State College of Ceramics, Alfred University, May 1973.
16. Segnit, E. R., and A. E. Holland: Formation of cordierite from clinochlore and kaolinite. J. Austral. Ceram. Soc. **7**, 43–46 (1971).
17. Tauber, E., and H. J. Pepplinkhouse: Ceramic properties of pyrophyllite from pambula, New South Wales. J. Austral. Ceram. Soc. **8**, 62–64 (1972).
18. Hedges, P. E.: Crystalline and glassy phases in commercially fired brick. Am. Ceram. Soc. Bull. **40**, 371–77 (1961).

19. Pauling, L.: Structure of complex ionic crystals. J. Am. Chem. Soc. **51**, 1010–26 (1929).
20. Brownell, W. E.: Subsolidus relations between mullite and iron oxide. J. Am. Ceram. Soc. **41**, 226–30 (1958).
21. Jander, W.: Reaction in the solid state at high temperature. I. Rate of reaction for an endothermic change. Z. anorg. allg. Chem. **163**, 1–30 (1927).
22. Jander, W.: Reaction in the solid state at high temperature. II. Reaction velocities of exothermic reactions. Z. anorg. allg. Chem. **166**, 31–52 (1927).
23. Ginstling, A. M., and B. I. Brounshtein: Diffusion kinetics of reactions in spherical particles. J. Appl. Chem. (U.S.S.R.), **23**, 1249 (1950).
24. Carter, R. E.: Kinetics model for solid-state reactions. J. Chem. Phys. **34**, 2010–15 (1961), **35**, 1137–38 (1961).
25. Hild, K., and G. Troemel: Die Reaktion von Calciumoxyd und Kieselsäure im festen Zustand. Z. anorg. allg. Chem. **215**, 333–44 (1933).
26. Jander, W., and E. Hoffmann: Reactions in the solid state at high temperatures. XI. Reaction between calcium oxyde and silicon dioxide. Z. anorg. allg. Chem. **218**, 211–23 (1934).
27. Brownell, W. E.: Crystalline phases in fired shale products. J. Am. Ceram. Soc. **33**, 309–13 (1950).
28. Brownell, W. E.: Reactions between alkaline-earth sulfates and cristobalite. J. Am. Ceram. Soc. **46**, 125–28 (1963).
29. Stern, K. H., and E. L. Weise: High Temperature Properties and Decomposition of Inorganic Salts. I. Sulfates, NSRDS-U.S. Nat. Bur. Stds., No. 7, October 1966.
30. Briner, E.: Calcul des énergies libres et des constantes d'équilibre des réactions de décomposition du sulfate de calcium seul ou en présence de silice. Helv. Chim. Acta **28**, 50–59 (1945).
31. Jang, S. D.: Solid-State Reactions Between Mullite and Alkaline-Earth Sulfates. M.S. Thesis, New York State College of Ceramics, Alfred University, June 1964.
32. North American Combustion Handbook. Cleveland, Ohio: The North American Manufacturing Company. 1965.
33. Brownell, W. E.: Black coring in structural clay products. J. Am. Ceram. Soc. **40**, 179–87 (1957).
34. Schoenlaub, R. A., W. Troyer, and K. Hoekstra: Burnout rates on a shale body. Am. Ceram. Soc. Bull. **45**, 257–59 (1966).
35. Osman, M. A., S. M. Ehmke, and J. F. Skelly: Core burnout in brick making. Am. Ceram. Soc. Bull. **49**, 193–200 (1970).
36. Houseman, J. E., and C. J. Koenig: Influence of kiln atmospheres in firing structural clay products: II, Color development and burnout. J. Am. Ceram. Soc. **54**, 82–89 (1971).
37. Brownell, W. E.: Efflorescence resulting from pyrite in clay raw materials. J. Am. Ceram. Soc. **41**, 261–66 (1958).
38. Blachère, J.: Desulfurization of pyrite. J. Am. Ceram. Soc. **49**, 590–93 (1966).
39. Lovejoy, R. J., J. H. Colwell, and G. D. Halsey: Infrared spectrum and thermodynamic properties of gaseous sulfur trioxide. J. Chem. Phys. **36**, 612–17 (1962).
40. Mellor, J. W.: A Comprehensive Treatise on Inorganic and Theoretical Chemistry, Vol. 10. New York: Longmans, Green and Co. 1930.
41. Remmey, F. B.: Kiln car top construction. Am. Ceram. Soc. Bull. **49**, 266–68 (1970).
42. Marshall, R. W.: Forced draft firing for beehive periodic kilns. Am. Ceram. Soc. Bull. **49**, 518–21 (1970).
43. Sakol, S. L., and I. S. Shah: Removal of sulfur dioxide from clay kiln exhaust gases. Am. Ceram. Soc. Bull. **49**, 317–19 (1970).
44. Hunter, O., Jr., and W. E. Brownell: Thermal expansion and elastic properties of two-phase ceramic bodies. J. Am. Ceram. Soc. **50**, 19–22 (1967).

45. Lachman, J. M., and J. O. Everhart: Development of safe cooling schedules for structural clay products. J. Am. Ceram. Soc. **49**, 30–38 (1956).
46. Edwards, J.: Thermal acoustical analyzer helps solve cooling problems. Brick and Clay Rec. **165**, 27–29 (1974).
47. Hasselman, D. P. H.: Thermal Stress Crack Stability and Propagation in Severe Thermal Environments. Ceramics in Severe Environments, W. W. Kriegel and H. Palmour, III, eds. New York: Plenum Press. 1971.

7. Decoration, Panels, and Packaging

7.1. Sanded Surfaces

Sand-coated surfaces are a type of decoration applied to face bricks, and they can be provided in nearly every color imaginable. The soft-mud process requires a coating of sand for forming, but this necessity is exploited to produce bricks of a variety of colors—very often copying the appearance of antique, hand-molded products. Sanded surfaces are also widely used on stiff-mud bricks, simulating the appearance of soft-mud bricks. This type of decoration, as well as all other types, can be produced in such a wide range of colors that chromatology will not be discussed extensively in this chapter. The colors produced by the control of the oxidation state of iron was discussed at length in Chapt. 6. Other transition-element oxides and previously prepared pigments are subject to the same basic principles. Our considerations of the decoration of structural clay products will be involved largely with textural appearances.

Many different types of sands and sand mixtures are used in the manufacture of soft-mud bricks to give the appearance shown earlier in Fig. 42. For soft-mud forming, sands as fine as minus 40-mesh are usually used, but some are produced with coarser grades—perhaps up to about minus 10-mesh. In any case a proper grain-size distribution is important for good mold release. Some of the sands are pure, such as white glass sands, but others are crude bank sands containing perhaps 50% to 60% quartz, the remainder of the minerals being unique to the particular deposit. Unadulterated, such sand coatings produce tan, brown, pink, and red colors on firing. Better coverage and a wider range of colors are produced by adding clays, coloring oxides, ceramic pigments, and fluxes such as soda ash and borax to the molding sands. Unevenness in brick shape, surface texture, and color application are often desirable to overcome the monotony of uniformity.

It is a bit more difficult to coat the exposed surfaces of stiff-mud bricks with these sands, due to the stiffness of the extruded column and its lack of stickiness. Usually the coatings are applied by sand blasting with compressed air or by rolling, to impress the sand grains into the surface for better adhesion. In an attempt to further copy the unevenness of the molded product, the rollers sometimes are equipped with a relief design to indent the smooth die surfaces while forcing the sand grains into the surfaces. Production equipment to perform this task on an extruded column is shown in Fig. 95. In this case the sands or powdered coatings are applied after the column is cut by a reel. An arrangement is made for four different sand mixtures to be available from overhead hoppers for coating the bricks. After leaving the sand-application nozzle, side rollers first fix the coating and impress an irregularity onto the ends of the bricks; then, a wider roller does the same for the major faces.

Fig. 95. Application of a sand or powder texture with irregular indentations, to the surfaces of extruded face bricks

Because there is no mold-release problem in sanding a stiff-mud column, grits with a narrow particle-size range can be employed for special effects. Natural sands, stained grains, previously prepared grog particles, and ground rocks can be sized from 8 mesh to 20 mesh or 10 mesh to 40 mesh for application. It is more common to add a flux to the sands for application to stiff-mud bricks in order to promote adherence.

7.2. Texturing of Extruded Bricks

Texturing the surfaces is a type of decoration unique to the extrusion method. These decorations would have to be classed as modern with respect to the long history of brickmaking, but this does not rule out the possibility of some ancient products being scratched or embossed for esthetic effects. It simply was not done on such an extensive scale by the old brickmakers.

Particle-size distribution of the basic raw materials from which the bricks are made plays a large role in all texturing of stiff-mud bricks. This is where the "texture fraction" referred to in Chapt. 3 comes into play. Smooth-surfaced bricks, tiles, and pipes should have no texture fraction as previously defined. All ingredients going into the production of such products should be less than 14-mesh to prevent surface hairline cracks which are apt to surround the coarser grains. In addition, some of the texturing operations also produce unique results when no texture fraction is present; however, many of the textures developed depend on a definite proportion of coarse particles.

The simplest of all textures is the wire- or knife-cut type. In this case the column is extruded deliberately oversize in order that 1/8 inch (3 mm) slices can be cut from the surfaces as the column leaves the die. The slicing operation is shown in Fig. 96

Fig. 96. Knife-cutting a texture on the surfaces of an extruded column

where the surface is peeled off and the scrap returned to the pug mill. As the knife passes through the column, the texture-fraction particles are dragged through the soft clay, thereby leaving scratches. Sometimes the particles being dragged are eventually pushed down into the surface of the column, and at other times they are pulled out. These actions leave at least two types of scratch on the surfaces of the bricks. If a very rough texture is desired, the texture fraction may contain particules as large as 2 1/2- to 3-mesh, and the results can be seen in Fig. 97. Finer particles in the texture fraction produce a more delicate wire-cut texture.

Fig. 97. A knife- or wire-cut texture on the surfaces of a face brick. Note the coarse particles added to provide this texture

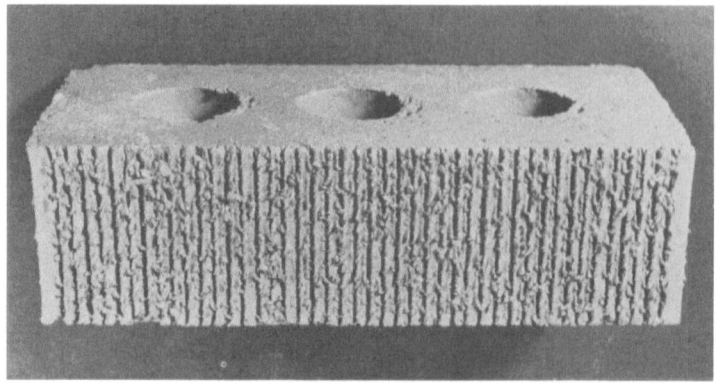

Fig. 98. Pin-scratched texture on an otherwise smooth face brick. Note the absence of coarse particles in this product

Fig. 99. Device for sweeping a random-scratched texture onto an extruded column of plastic clay

Many other devices to apply textures to a smooth column have been invented by engineers in the brick factories, and probably many more will be developed as engineers are asked to provide special effects for their customers. Fig. 98 shows a texture resulting from scratches made by fixed pins or needles. In this case little or no texture fraction is used in the particle-size grading of the raw materials. To produce a more random scratched texture a device like that shown in Fig. 99 can be used where loose wires are rotated rapidly to "sweep" a texture onto the surface of the column. This machine creates a texture such as that on the brick in Fig. 100.

All of the textures shown so far will give a vertical mode when the bricks are placed in a wall. It is more difficult to apply a horizontal texture to such a moving

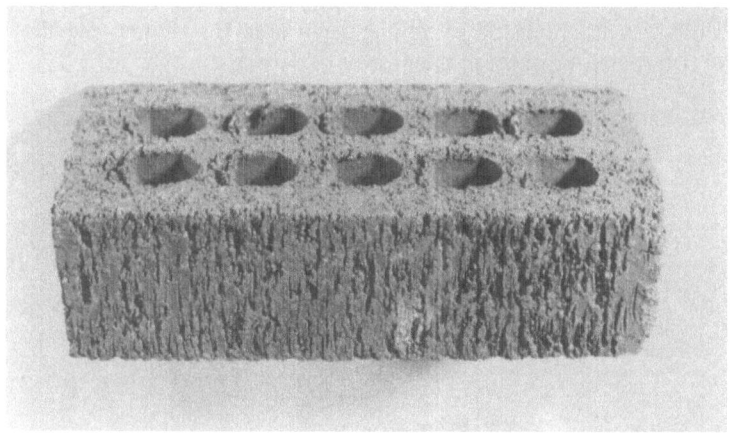

Fig. 100. Face brick with a scratched texture, produced by the machine shown in Fig. 99

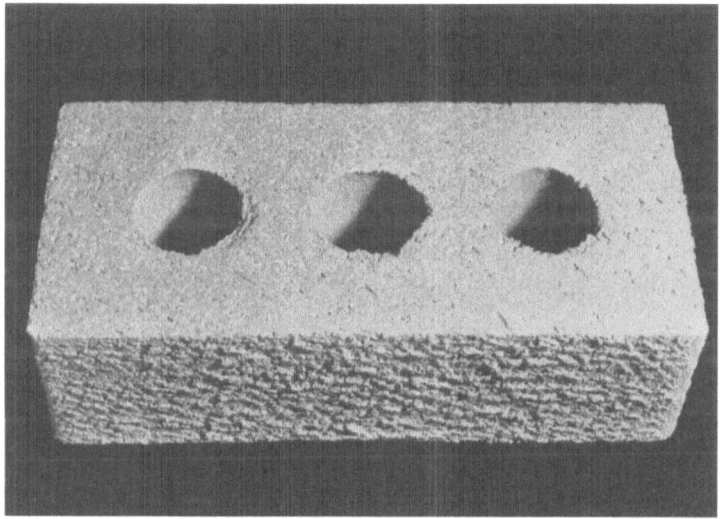

Fig. 101. Face brick with a roughened surface, on which horizontal lines predominate

column; however, devices that sometimes resemble a reel-type lawn mower are used to give a horizontal motif. Such a texture can be seen in Fig. 101.

Quite different textures can be imposed on the smooth surfaces of face bricks by means of rollers equipped with relief patterns. Some thought is given here to the size of the roll and to the speed of the column so that each brick cut from the column is not exactly like the next. The equipment for rolling on a texture may look like that in Fig. 102, and the final results can be of unlimited varieties; two are shown in Figs. 103 and 104. The imprints of Fig. 104 were originally made from plaster casts of real leaves, ferns, twigs, etc. The rounded edges where the column was cut into bricks by wires was accomplished by rolling paper onto the column and allowing the wires to cut through the paper in the process of cutting the column.

Fig. 102. Device for rolling a texture onto a smooth column

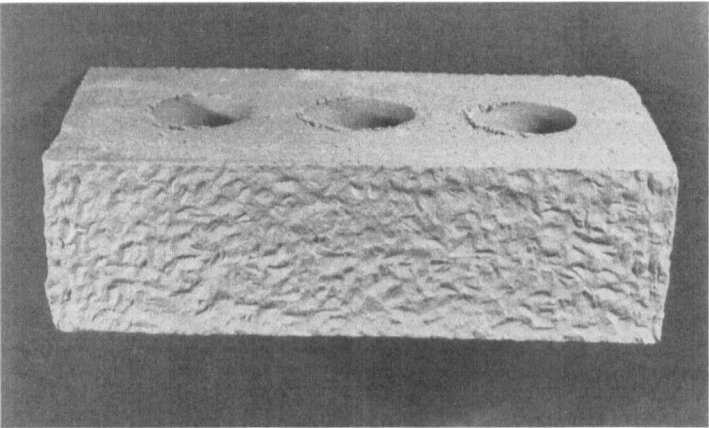

Fig. 103. A relatively fine texture, impressed on a face brick by a roller similar to that of Fig. 102

The imposition of more formal designs onto wall and floor tiles is done by the ram used in press-forming.

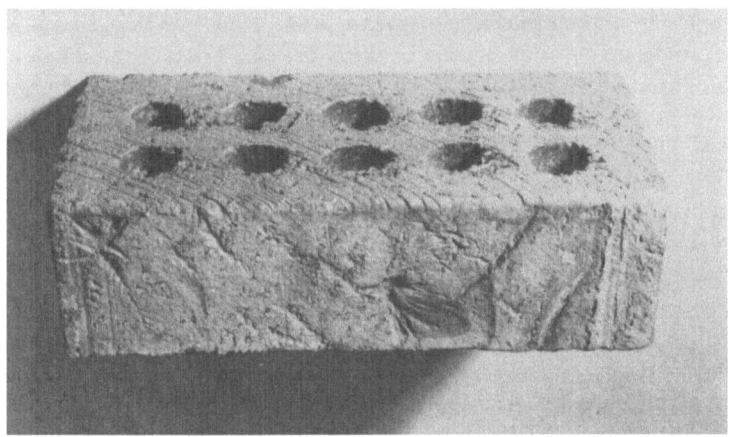

Fig. 104. Face-brick texture with a natural motif applied by a roller

7.3. Coating Decorations

Engobe and glaze coatings are used on floor tiles, wall tiles, and face bricks. In the face-brick industry, engobe-like materials are usually called slurries, probably because of the wide variety of application techniques used, some of which are deliberately not smooth and are of nonuniform thickness. Both engobes and glazes are applied in a thick, paint-like consistency, and as such they are called *slips. Engobes* and *slurries* are simply extensions of the body, usually designed to change its color or texture; however, *glazes* are essentially glassy coatings applied to a largely crystalline body.

7.3.1. Engobes and Slurries

Since engobes and slurries are applied in the slip condition to a plastic or dry body, they must be prepared from selected raw materials to be compatible with the body in all respects. The coatings should shrink with the body during drying and firing to prevent them from cracking off, and the thermal expansions must match closely enough so that cracking does not occur on cooling. On firing, engobes must mature to a stable product with absorptions and densities similar to the body. As such, the coatings should not become sticky during firing, and cross or face settings of bricks should present no problem.

Engobes are usually designed to change the color of the body; therefore, they must be prepared from materials that will give the desired color. Usually, they are prepared from white-firing materials to which colorants can be added as desired. Some clay is essential to engobes in order to produce stable slips that do not settle out immediately. Some clays that can be used are china clays, plastic kaolins, or ball clays. In order to control shrinkage and provide some fluxing for these refractory clays, such materials as feldspar, nephelene syenite, aplite, flint, and alumina can be added. A typical composition of an engobe or slurry for face bricks is given below, to which lead bisilicate can be added in any desirable amount as a low temperature flux [1].

45% potter's flint

17% nephelene syenite

23% ball clay

9% talc

6% alumina (corundum)

In the face-brick industry, slurries are applied by many different techniques, but when a uniform coating is desired, spray-guns are used. For good application the slips should be deflocculated with a chemical such as sodium hexametaphosphate, to provide some thixotropy and to keep the water content as low as possible. (*Thixotropy* is that property of a slip that appears as gelation or increased viscosity when stirring or agitation ceases.) The slips are made up to specific gravities between 1.75 and 1.84, and a satisfactory spraying consistency is achieved with specific gravities from 1.75 to 1.78 [2]. Engobes are also applied with paint rollers, spatulas and brushes, to achieve texture effects as well as to coat. In these application techniques, the specific gravities are higher than those recommended for spraying. Sometimes engobes are applied over a pretextured product like the brick shown in Fig. 105.

Fig. 105. A textured face brick with an engobe coating

Another reason for the application of an engobe to a structural clay product is to provide a better fit for a subsequent glaze application. In this case the engobe acts as a body intermediate between the product and the glaze as far as thermal expansion is concerned. At times an engobe assists the covering power of a particularly desirable glaze.

7.3.2. Glazes

Glazes must be considered only as a decoration on structural clay products. They should never be applied for the primary purpose of providing a water-tight seal, greater strength, or an easy-to-clean surface. Since these products are all exposed to moisture, and since bricks especially are subjected to severe weathering, the primary consideration should be durability of the glazed product. Let us not forget the tragedy of the terra cotta industry narrated in Chapt. 1. At this time there are no generally accepted specifications for glazed facing bricks, but experience has shown that the product should meet the severe weather specifications of the American Society for Testing and Materials before the application of a decorative glaze is considered. In this regard, it would be a mistake to accept the strength waiver over absorption and saturation coefficient that is allowed in these specifications. (More details of the durability specifications are to be found in Chapt. 11.) When glazes are applied, no compromise can be made with firing temperature to achieve a particularly attractive glaze [1].

Glazes are fundamentally different materials from the bodies on which they are attached; therefore, special attention must be given to compatibility. Since glazes are glass films, for all intents and purposes, they are very brittle materials that crack easily under tension stresses. For durability as well as appearance, glazed surfaces should not be cracked, or *crazed*; therefore, the body must place the glaze in compression during cooling, by means of a small differential in thermal expansions. Before designing a glaze for structural clay products, the coefficient of linear thermal expansion of the mature body on which it is to be applied should be measured. It will be found that the coefficients for bricks fall into the range of 57×10^{-7} to 75×10^{-7} cm/cm/°C. Probably most floor and wall tile bodies will give coefficients of expansion closer to the higher value [1].

In order that the glaze will develop a compression stress on cooling, it must be designed to have a coefficient of linear thermal expansion about 10×10^{-7} cm/cm/°C lower than the body to which it is applied; however, somewhat greater differences can be tolerated before the glaze fractures in shear, causing *shivering*. An estimation of the thermal expansion of a particular base glaze can be calculated by empirically derived factors for the oxides present. Table 14 lists factors, derived from several sources, for oxides commonly found in glaze compositions. To calculate the linear expansion coefficient for a glaze, the percentage of each oxide is multiplied by its factor, and the products of multiplication are added to obtain the coefficient. It is generally safe to exclude opacifiers and colorants from this calculation, because their effects will be to lower the thermal expansion. This is going in the right direction to obtain a proper fit, since there is usually a wide latitude in compressive stress before shivering occurs. The factors for the various oxides, make it evident that it will be more difficult to produce a low-temperature glaze having a low expansion.

Since the raw materials of glaze batches can vary widely, the mole-fraction glaze formulas for several successful glazes will be reported here only as typical examples. Ground glasses, known as *frits*, are available in many compositions, and these are common raw materials for glazes; however, for some applications raw glazes can be prepared that contain no premelted frits. It must be remembered that all glazes must

Table 14. *Factors for Calculating Co-*
efficients of Linear Ther-
mal Expansion of Glazes
(cm/cm/°C)

Glaze Oxide	Factor $\times 10^{-7}$
SiO_2	0.27
Al_2O_3	0.17
B_2O_3	0.03
Na_2O	4.32
K_2O	2.83
CaO	1.67
MgO	0.45
PbO	1.00
ZnO	0.60
BaO	1.73

contain from 3% to 15% clay, depending on its particle size, to provide a stable, slightly thixotropic suspension. An opacified glaze formula maturing at 2150°F (1177°C) and having a coefficient of linear thermal expansion of 47×10^{-7} cm/cm/°C is [3]

$$\left.\begin{array}{l} 0.05 \ Na_2O \\ 0.12 \ K_2O \\ 0.63 \ CaO \\ 0.20 \ ZnO \end{array}\right\} \quad 0.35 \ Al_2O_3 \quad \left\{\begin{array}{l} 3.14 \ SiO_2 \\ 0.35 \ ZrO_2 \end{array}\right.$$

A 1900°F (1038°C) glaze with a coefficient of linear thermal expansion of 62×10^{-7} cm/cm/°C has the formula

$$\left.\begin{array}{l} 0.03 \ Na_2O \\ 0.01 \ K_2O \\ 0.05 \ MgO \\ 0.16 \ CaO \\ 0.75 \ PbO \end{array}\right\} \quad \begin{array}{l} 0.25 \ Al_2O_3 \\ 0.04 \ B_2O_3 \end{array} \quad \left\{\begin{array}{l} 3.03 \ SiO_2 \\ 0.21 \ ZrO_2 \end{array}\right.$$

An opacified glaze which matures at 1800°F (982°C) and has a coefficient of linear thermal expansion of 61×10^{-7} cm/cm/°C is

$$\left.\begin{array}{l} 0.02 \ Na_2O \\ 0.04 \ CaO \\ 0.94 \ PbO \end{array}\right\} \quad \begin{array}{l} 0.19 \ Al_2O_3 \\ 0.12 \ B_2O_3 \end{array} \quad \left\{\begin{array}{l} 3.24 \ SiO_2 \\ 0.21 \ ZrO_2 \end{array}\right.$$

Finally, a low-temperature glaze, especially designed for the brilliant selenium-cadmium sulfide stains, is presented. This glaze matures at about 1500°F (816°C) and has a coefficient of linear thermal expansion of 76×10^{-7} cm/cm/°C.

$$0.30 \text{ Na}_2\text{O} \left.\begin{array}{l} \\ 0.22 \text{ K}_2\text{O} \\ 0.05 \text{ ZnO} \\ 0.28 \text{ PbO} \\ 0.15 \text{ CdO} \end{array}\right\} \quad \left.\begin{array}{l} 0.12 \text{ Al}_2\text{O}_3 \\ 0.71 \text{ B}_2\text{O}_3 \end{array}\right\} \quad \left\{2.64 \text{ SiO}_2\right.$$

It will be noticed that the maturing temperatures of these glazes suggest one-fire and two-fire types. If the glaze matures at the same temperature as the body, it can be applied to the green ware, and a glazed product results from a single firing operation. Most structural clay products are made this way, but special market demands often require the more expensive two-fire procedure. Of course, in the glazing of bricks they must be set flat on the kiln cars to prevent them from fusing together.

One final thought on the selection of clays for glazes is that the glaze materials must shrink with the body until fusion starts, to prevent cracking of the glaze, which will result in *crawling*. To assist in the adherence of the glaze coating during drying and the early stage of firing, binders such as methocel, gum arabic, or lignosulfonates are often added to the glaze batch.

7.4. Panelling

For simplicity of installation, factory construction of sheets or *panels* of floor and wall tiles has been common practice for many years, but a similar prefabrication of brick walls is just beginning to find an acceptable market. Some floor and wall tiles are glued in their proper order to sheets of paper for quick installation; others are now grouted with finished joints made of a flexible, durable, latex-type material. These sheets need only to be glued to the wall surfaces. Prefabricated walls of brick are now bonded with excellent mortar joints made from a quick-setting, high-early-strength cement.

The greatest advantage of brick panelization to the structural clay products industry is closer control over the use of their products. Up to this time, recommendations for the best masonry practices for particular bricks have been made to builders, but beyond this the brick manufacturer has had little control over the end-use of his product. At times the brick maker has been blamed for the results of poor construction practices. The factory-produced panels have stronger and more uniform bonds between mortar and brick than has been possible with hand-laying procedures. It is also quite possible that factory-prefabricated walls could be placed in the wall of a building more economically.

Consumer resistance to brick panels may result from the size and shape uniformity that must be a part of the panelizing processes. The construction of buildings would have to be standardized in order to make universal use of factory-made panels. This would be a limiting factor to the creativity of some architects, but perhaps the more reliable construction will induce some favorable compromise. Further development work on panelling of face bricks is necessary to make such units acceptable to the home-building market.

7.5. Packaging

Packaging of structural clay products has proven to be an economic advantage to all concerned with the construction industry. In the factories, it has reduced losses induced by handling, and has provided for an efficient inventory. One of the greatest advantages to result from packaging is the reduction of breakage and soiling during shipment. Packaging has been designed to allow for easy handling during loading, unloading, and on the job-site.

Floor and wall tiles have been packaged conveniently for many years, but the packaging of bricks and pipes has come into common practice only during the last twenty years. The Structural Clay Products Research Foundation was largely responsible for the trend toward the packaging of bricks, and the National Clay Pipe Institute provided the developmental research necessary for the packaging of sewer pipe. Improvements in packaging are constantly being made at various factories, as better ideas and materials are forthcoming.

In the brick industry, packaging has become an automated process which is just now being integrated into a totally automated system by devices for automatic unloading of kiln cars. Obviously, such a system can only be attempted where the controls on forming, drying, and firing are so good that practically no inspection is required at the unloading point. Fig. 106 shows a conventional strapping machine and a package of bricks emerging. Note the channels near the bottom of the package for use by fork-lift trucks. Many of these packages are strapped so that the large

Fig. 106. Automatic packaging machine and an emerging package of face bricks

shipping package can be reduced to smaller packages on the job. In many cases paper is placed between the layers of bricks to prevent chipping, which becomes extremely important for coated and glazed products. Glazed bricks may be further protected by paper and sheets of plastic.

Fig. 107. Sewer pipes vacuum-packed in sheet plastic

The strapping of sewer pipes on wooden pallets is common packing practice, as it is for face bricks. Such a package can be seen in Fig. 111. A more recent idea in the packaging of sewer pipes is shown in Fig. 107. These pipes are vacuum-packed in sheet plastic that is strong enough to make the usual strapping unnecessary. This type of package keeps the products clean regardless of shipping and storage conditions, and it further protects the plastic joints from damage.

References

1. Groskaufmanis, E.: Ceramic glazes on structural clay materials. J. Can. Ceram. Soc. 27, 108—11 (1958).
2. Thomas, D. W.: Preparation and application of engobes to brick. Brick Clay Rec. 142, 58—62 (1963).
3. Jacobs, C. W. F.: Glazes for brick and structural clay products. Ceramic News 13, 9, 16 (1964).

8. Jointing of Vitrified Clay Sewer Pipe [*]

8.1. Factory Installed Jointing Units

When the first sewer collection systems were installed, leak-proof sewer lines were not considered to be necessary. As a matter of fact, leaking lines were often considered to be desirable because in low areas leaking joints contributed to the lowering of the ground-water level. Little thought was given to the exfiltration of sewer waters into the ground, since pure-water systems were being installed in urban areas at about the same time as sewer systems.

The first-used methods of jointing were later found to be unsatisfactory for several reasons. When the jointing of pipe was done by the workman in the ditch, using cement mortar tamped into the annular space between the spigot end and the bell cavity of the adjoining pipe, leaky joints were common, durability of the jointing material was poor, and the rigidity of the joints promoted cracking due to ground pressures of all kinds. The pouring of hot asphaltic and coal-tar compounds into the joints was an improvement in durability and flexibility, but leaking joints persisted because the working conditions did not foster good workmanship. In subsequent years prior to World War II, a myriad of other methods was developed for jointing pipe. The most common of these were molded rubber gaskets which were placed over the spigot end of the pipe, then forced or pushed into the bell-end of the adjacent pipe. All of these methods were beset with the problem of quality control of workmanship and with the element of human error.

As in all of the structural clay products industries, the sewer-pipe manufacturers have had little control over the end use of their products. Over the years, this has turned out to be a major problem in serviceability. The sewer-pipe industry has made significant progress with respect to this problem by employing factory-installed jointing systems that require a minimum of effort and skill on installation, to achieve durable, tight, and flexible joints.

8.2. Requirements for Good Joints

The basic requirement for all jointing systems for vitrified clay pipe is a leak-proof assembly, and this demands precise dimensional control of the outside diameter of

---

[*] A. J. Reed, President, National Clay Pipe Institute, assisted in the preparation of this chapter.

the spigot end and the inside diameter of the bell-end of the pipe. As in the case of most ceramic materials, there is appreciable shrinkage in drying and firing of the pipe. As a result, it is necessary to apply another material to the inner surface of the bell and outer surface of the spigot, in order to obtain precise dimensional control. The best leak-proof joints are made by employing the compression of a resilient or elastomeric material, such as rubber, between two dimensionally true surfaces, to a known and reproducible degree of compression. Since this is not possible between the clay surfaces of sewer pipe, a great deal of research has been conducted, over the years, to develop resins for casting on the pipe-ends and molds to provide the precise, dimensionally controlled surfaces. Then, a compression member of the joint is designed to always give a tight joint by controlling the degree of compression.

In addition to being leak proof, both with regard to infiltration and exfiltration, there are other necessary requirements for pipe joints. These are:

1. The joint must be easily assembled by unskilled laborers.

2. The joint must be assembled without the use of excessive force to push one end into the other.

3. The joint must be flexible, so that the sewer-line can adjust to earth movements.

4. The joint must resist differential loading by the soil backfill in the event that the barrel of the pipe is improperly bedded on the trench bottom.

5. The materials used in truing the jointing surfaces and in the compression member must be resistant to the chemicals often found in sanitary sewer systems.

6. The joint must be tight, to resist penetration by roots.

8.3. Types of Compression Joints

In the past 30 years, there has been a continuing program of research in the clay pipe industry, in order to develop new and improved materials and designs for jointing of pipe. Although several good systems of factory-installed jointing units are being employed, the state of the art, at this time, is still in a state of flux. The optimum jointing system with regard to both dependability in service and economics has not yet been established; therefore, we find various systems being used—sometimes more than one in a single factory.

8.3.1. Polyvinyl Chloride Joints

The first material used for compression-type, resilient joints was plasticized polyvinyl chloride (PVC), and it has many good properties. The system was developed about 1950 and is still in limited use today. This material has the basic requirements for a good jointing material, and it is liquid prior to curing, so that it is castable; but it requires a relatively high temperature ($350°F$; $177°C$) to cure the polymer to a predictable and controlled flexibility. The curing process is more appropriate for small pipes, due to the time and care required to heat and cool large ones.

Vinyl chloride has the formula, $CH_2 : CHCl$, and it polymerizes when a catalyst, or initiator, breaks the double-bond producing a variable-length chain as shown below [1].

$$
\begin{array}{cccc}
H & H & H & H \\
| & | & | & | \\
-C- & C- & C- & C- \\
| & | & | & | \\
H & Cl & H & Cl
\end{array}
$$

The factory receives a low-molecular-weight polymerized vinyl chloride in the form of a thick, syrupy liquid to which a little plasticizer, catalyst, and chain regulator probably have been added. *Plasticizers* are large organic molecules which tend to separate polymerized chains, and have no other chemical function in the polymerization process. Some plasticizers are tricresyl phosphate $[C_6H_4(CH_3)_3PO_4]$, dibutyl phthalate $[C_6H_4(COOC_4H_9)_2]$, dioctyl phthalate $[C_6H_4(COO(CH_2)_7CH_3)_2]$, polymeric esters and nitrile rubber [1]. The *catalyst* serves to extend the polymerization, and a mercaptan-chain-transfer *regulator* provides the desired chain length and cross linking. Some commonly used catalysts are dibenzoyl peroxide $[(C_6H_5)COO(C_6H_5)COO]$, dibutyl peroxide $[(C_4H_9)COO(C_4H_9)COO]$, dicaproyl peroxide, diazoamino compounds, diazothio ethers, ammonium perchlorate, metal persulfates, and sodium perborate. A mercaptan for the control of free radicals might be benzyl mercaptan, $C_6H_5CH_2SH$, or some similar compound.

The pipe joint is made by pouring the catalyzed polyvinyl chloride liquid into precision molds at both ends of the pipe, and heat-treating the assembly in an oven at 350°F (177°C) until the polymerization process gives the desired properties to the elastomer. When cured, PVC is a semirigid plastic with a high tensile strength and a low compression set. In the bell-and-spigot type joint, it is not necessary to use a separate compression member, since the jointing surfaces are dimensionally true and

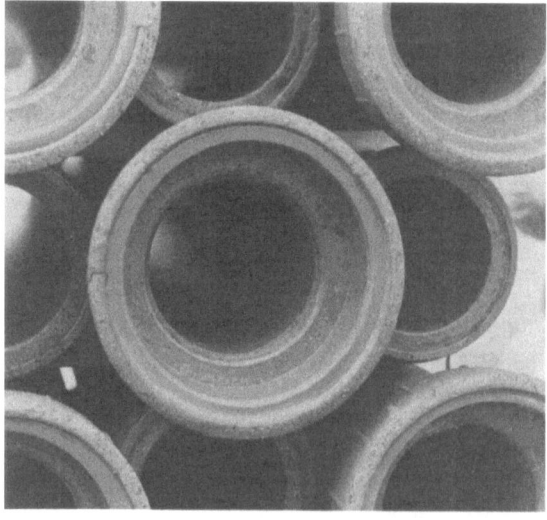

Fig. 108. Polyvinyl chloride jointing units attached to bell-and-spigot sewer pipes

sufficiently elastomeric to provide the required compression. The finished plastic joint is resistant to light, moisture, flexing, and to most chemicals. It withstands strong acids and alkalies as well as nearly all inorganic compounds. The PVC is not resistant to ketones, chlorine, nitro compounds, or gasoline [1]. Sewer pipes equipped with this type of plastic for jointing are shown in Figs. 108 and 109. The spigot end can be seen close up in Fig. 109.

Fig. 109. Polyvinyl chloride jointing unit, attached to the spigot end of a sewer pipe

8.3.2. Polyester Joints

The driving force for the development of polyester resin joints was the elimination of the oven-curing step. A semiflexible polyester resin using catalysts and promoters can be cured at room temperature to a sufficient degree of hardness, where the cast iron molds can be removed a few minutes after casting. The casting of the polyester mixture into the mold at the spigot end is shown in Fig. 110. A polyester resin does not have the elasticity of polyvinyl chloride; therefore, the polyester is used only as a truing material to obtain dimensionally accurate surfaces. It is, then, necessary to use a separate compression member, such as a rubber ring, which is usually placed on the spigot of the pipe and compressed against the bell when the joint is made.

As might be expected from the name, polyester resins are prepared from an unsaturated polyfunctional acid such as maleic acid, and a polyfunctional alcohol, such as diethylene glycol, by heating in an inert gas. The resultant polymer is blended with an acrylate or styrene monomer, to achieve the desired consistency, and an inhibitor to increase shelf life. A typical equation for this preparative reaction is given below. The added monomer will eventually become a part of the final polymer; therefore, its selection is important, because the final properties of the polymer will be affected by it. For example, ethyl acrylate will yield a more rubbery polymer than styrene [1].

Fig. 110. Casting a polyester jointing unit on the spigot end of a sewer pipe

$$\begin{matrix} O & & O & & O \\ \diagdown & & \diagup \diagdown & & \diagup \\ C & & C & \\ | & & | & \\ HC & = & CH & \end{matrix} + HOCH_2CH_2OH \xrightarrow{\Delta} (-O-CH_2CH_2-O-\overset{O}{\overset{\|}{C}}-CH=CH-\overset{O}{\overset{\|}{C}}-O-)_n$$

maleic anhydride	ethylene glycol	polyester M.W. 1500 to 3000

$$(-O-CH_2CH_2O-\overset{O}{\overset{\|}{C}}-\overset{O}{\overset{|}{CH}}-CH-\overset{O}{\overset{\|}{C}}-O-)_n$$
$$X-C-H$$
$$H-C-Y$$
$$(-O-CH_2-CH_2-OC-CH-CH-\underset{O}{\overset{\|}{C}}-O-)_n$$

cross-linking agent and catalyst

The cross-linking agents may be styrene, diallyl phthalate, or vinyl acetate. As an example of the cross-linking process, styrene has the formula, $C_6H_5CH:CH_2$; then, in the previous equation for the preparation of cross-linked polyester, the X could be C_6H_5 and the Y could be H.

When it comes time to use the polyester in the plant, an initiator, promotor, and filler are added to it and mixed thoroughly. A promotor is necessary to control the production of free radicals which determine the rate of polymerization, because the oxidizing initiators, previously discussed in connection with polyvinyl chloride, are

Fig. 111. Polyester jointing units on bell-and-spigot sewer pipes

Fig. 112. Polyester jointing unit attached to the spigot end of a sewer pipe

difficult to control. Bulk fillers are used in this application to increase the resin volume and to lower the cost. They may be calcium carbonate, clay, other silicates, or similar materials.

Figs. 111 and 112 show the polyester resin applied to the ends of bell and spigot pipes. Notice the groove in the plastic on the spigot end of the pipe in Fig. 112 where the rubber ring is placed to act as the compression member of the joint. The durability of the polyester-rubber joints is as satisfactory as those of polyvinyl chloride.

8.3.3. Polyurethane Joints

Next, in the development of plastic joints for sewer pipes, was the use of polyurethane which requires only moderate heat for curing but still gives sufficient elasticity to effect the necessary compression without the use of a separate rubber ring. When cured, polyurethane has much the same physical properties as PVC, but it is formulated to cure at a higher degree of hardness. The principal design used with polyurethane is known as the "bead compression joint" where an integral bead is molded in the plastic on either the spigot or bell-casting. The bead acts as the compression member between the jointing surfaces. The bead varies in dimension, depending on the hardness of the polyurethane and pipe size, but it is approximately 0.05 in. to 0.12 in. (0.13 to 0.30 cm) high. The polyurethane joint is in extensive use throughout the industry on all sizes of pipes from 4 in. (10 cm) through 42 in. (107 cm) in diameter. In many areas it has replaced the PVC joint.

Cross-linked molecular polyurethanes are reaction products of polyhydroxy compounds and polyisocyanates where one monomer is difunctional and the other trifunctional. Ethylene glycol could be the difunctional monomer and some compound similar to toluene-2, 4-diisocyanate could serve as the trifunctional component. The structure is

$$CH_3$$

— NCO

NCO

When a hydroxyl-rich, low-molecular-weight polyester is provided, a flexible elastomer can be produced by adding excess diisocyanate to lengthen the chains, glycols, diamines, consistency modifiers, and fillers and heating ("vulcanizing") at 140°F (60°C). A common diisocyanate is naphthalene diisocyanate which has the formula

— NCO
— NCO

The ethylene glycol formula was given in the equation for the preparation of polyester. A commonly used diamine is hexamethylenediamine, $H_2N(CH_2)_6NH_2$. Water must be excluded from these reactions, or some unwanted expansion of the plastic will take place due to reaction with the isocyanate.

Because there are so many variations in composition of the three types of plastics used in jointing, and because each variation has its own properties, only the general chemistry relevant to them has been discussed. Specifics have been avoided because the optimum material and best process probably have not been achieved, and changes in the chemistry of jointing materials may go on for some time. When one considers the polyesters alone as being complicated organic salts created from the reactions between organic acids and alcohols, one intuitively feels that there must be almost an infinite number of polyester resins, each with its own properties. Polyvinyl chlorides are modified with initiators, chain-length and cross-linking regulators, and plasticizers. The possible combinations of compounds which can produce polyurethanes are almost as extensive as for polyesters. In the factory, a general understanding of the materials and processes is probably all that is necessary to keep quality under control and to appreciate modifications that may be proposed from time to time.

8.3.4. Jointing of Plain-End Pipes

In recent years the sewer-pipe industry has been producing plain-end pipes that are simpler in form and easier to produce. These pipes are possible because jointing systems have been perfected that eliminate the need for the typical bell-and-spigot connection. Jointing is accomplished either through the use of a rubber or plastic collar, or through the use of a low-profile, resin-type bell which is attached to one end of the pipe. There are many versions of these jointing systems in use today.

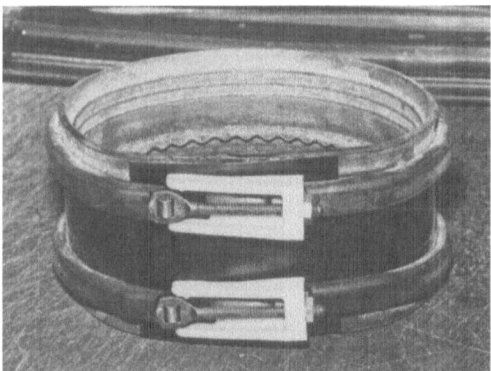

Fig. 113. Collar unit for jointing plain-end pipes

A collar unit for jointing plain-end pipes is shown in Fig. 113. The butyl-rubber collar contains two stainless-steel bands that are tightened by silicon-bronze bolts and nuts, supported by a rigid plastic bridge to effect the necessary compression. The collar is placed on the ends of the pipes to be jointed as shown in Fig. 114 to achieve a tight, flexible connection. These collars are shipped with the pipes and are installed on the job.

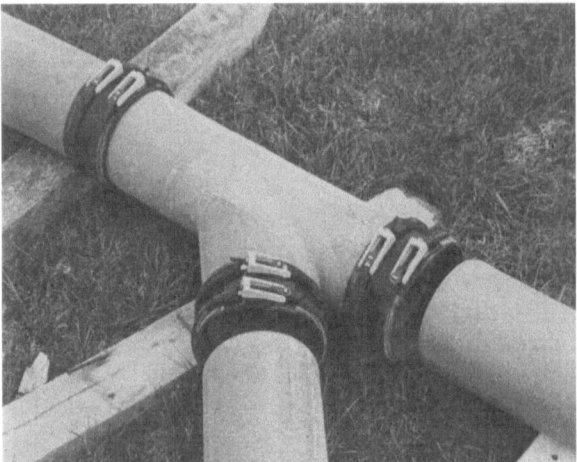

Fig. 114. Plain-end sewer pipes jointing by butyl-rubber collars

In the production of plastic low-profile bell-and-spigot joints for plain-end pipes, polyurethane or polyester castings in the form of a bead or ridge are placed on both ends of the pipe. A rigid, polyvinyl-chloride sleeve, reinforced with a stainless-steel ring, is forced onto one end of the pipe, providing the prearranged compression. This type of jointing unit is shown in Fig. 115. The other, beaded pipe end is forced into the slightly flared sleeve to make the joint in the field.

A variation on the low-profile bell design is the use of a monofilament glass-reinforced polyester. In this modification, a bell is wound on one end of the pipe

Fig. 115. Polyvinyl chloride-steel sleeves, attached to polyurethane beaded plain-end pipes to effect a compression joint

by passing a monofilament glass through a polyester resin bath and helically winding the glass on the end of the pipe and a projecting mandrel. The adjacent spigot in the jointing system may then have either a rubber-gasketed polyester resin casting or a beaded polyurethane casting, to provide the compression.

Over the years, millions of dollars have been spent by the clay-pipe industry on research to improve the jointing systems. Grinding the fired spigot-and-bell pipes, in order to produce dimensionally accurate surfaces, is one of the research efforts. With this technique, only a compression member would be required between the two dimensionally true surfaces to provide the desired compression.

8.4. Specifications and Tests for Vitrified Clay Pipe Joints

The specific chemical and physical properties required of sewer-pipe joints are outlined in the American Society for Testing and Materials specifications. These currently are C594-74, "Compression Couplings for Vitrified Clay Plain-end Pipe," and C425-74, "Compression Joints for Vitrified Clay Bell and Spigot Pipe." Included in test requirements are standards for chemical resistance, tensile strength, hardness, compression set, and ozone resistance.

The completed joint is also required to meet certain performance standards as outlined in the applicable ASTM specifications. In the field, exfiltration or infiltration tests are usually conducted; although, low-pressure air tests are presently increasing in use to determine the tightness of the completed sewer line. ASTM requires that certain performance tests be met which relate to the end-use of the jointed pipes in the ground. Among these are differential loading (commonly known as shear loading), deflection of bell-and-spigot joints and some types of plain-end joints, and displacement in some types of plain-end joints. Independent specifying agencies or code groups may require certain performance tests other than those outlined in the ASTM specifications.

References

1. Winding, C. C., and G. D. Hiatt: Polymeric Materials. New York: McGraw-Hill Book Company. 1961.

9. Quality Control

9.1. Philosophy

Effective, organized quality control in any factory is partly philosophical and partly technical. The pressures on the superintendent to lower unit costs tend to cause relaxing of quality standards by expediting one or more of the process steps. It has been proven over and over again that it is foolish economy to increase production at the expense of quality. Supportive cooperation between production personnel and those of quality control can only be achieved when each has parallel responsibilities stemming from an overall plant manager who is responsible for the long-range, economic health of the company. Once the human relations are firmly established for the inclusion of a quality-control program, optimum profits result from reduced production losses, ware uniformity that facilitates automated production, and high-quality products that overcome customer complaints [1].

It is time to review the previous chapters in order to pick out significant points that affect product quality and production losses. Raw-material uniformity or, at least, the recognition of variations, becomes a controlling factor throughout all the subsequent operations. The mineralogy affects forming, drying, and especially firing. Particle-size distribution, with particular attention to the three fractions, has been shown to relate to plasticity, decoration, and drying. The development of maximum plasticity for the forming of structural clay products is an extremely desirable condition, and at the time of forming, plastic strength is controlled by the amount of water added. When various coatings and glazes are used for esthetic purposes, a side-line or parallel quality-control program will be necessary. The various processes mentioned extend from acquisition of raw materials to forming of the product.

In modern factories the important factors in the remaining processes, namely drying and firing, are preset, automatically controlled, and usually require only superficial monitoring. In any case, records are usually kept which can be used to tie these processes into the whole quality-control program. During drying, the important variables are temperature, relative humidity, and time; whereas, firing is determined by temperature, time, and atmosphere.

Finally, the finished products must be sampled and inspected for durability criteria, visual appearance, dimensional tolerances, and losses due to distortion and cracking. This is probably the most important part of quality control, because all other attempts to keep the various processes within acceptable limits is for the purpose of producing acceptable products with minimum losses.

9.2. Nature of the Quality-Control Program

Experience has shown that quality control is evolutionary, from an initially complex program to a simpler system; however, since the needs of each plant are apt to be different, there seems to be no way of short-circuiting this evolutionary process. At the outset little is known about the variations that actually exist, so all of the data acquired are important; but as information is collected, certain elements of quality control will be found to be dependent on others. Some tests can be practically eliminated while others can be made at less frequent intervals. In addition, it is possible that experience in the program will suggest new, more pertinent testing procedures.

The target of any quality-control program is establishment of uniform quality in the final products, by checking the manufacturing process at appropriate points and making any necessary corrections. The particular control points require considerable thought. Some points in the process are more critical than others for their ultimate effects on the final product. Tests must be selected or devised to produce information quickly enough to allow the necessary corrections. Reliability of sampling and testing is absolutely essential, since bad data can cause changes to be made in the wrong direction.

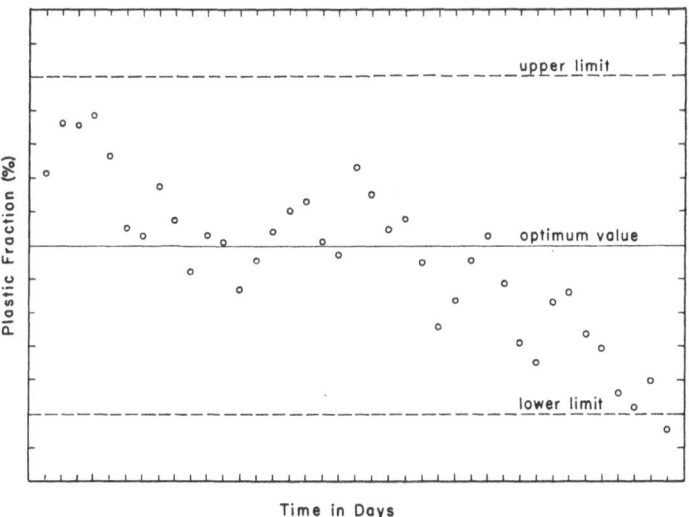

Fig. 116. Typical quality-control plot for a process variable tested on a daily basis

Permissible limits on variations that may occur within any process are unknown at first, but after some correlation between final results and the variations encountered, limits can be established. Once limits are established, charts like that of Fig. 116 can be used to anticipate when a process will run out of control and some correction is necessary to keep the values within the acceptable limits. Sometimes limits on the properties of the final product may be prescribed by standard specifications; so they will be given at the outset.

The collection of data on the critical points of production is really more important when things are going well than when trouble arises. If facts are not available when a problem becomes evident, e.g., breakage during firing, each process foreman blames the preceding one until the argument ends up in the mine where no one can confirm or deny that a variation has occurred to cause the problem. As a matter of fact, a significant change in the raw materials mined from a specific deposit is very unlikely; therefore, the raw materials may be the least likely place for trouble to start.

9.3. Procedure

If the raw materials have been properly evaluated *in situ*, there is often no need for further monitoring of their uniformity except for special situations. In cases where blending of two or more raw materials is made, it is sometimes necessary to check the proportions for the proper development of fired color a few days before the material enters the system. This is done by preparing test specimens, firing them under plant conditions, and making color comparisons with previously established standards.

When a deposit contains soluble sulfates that cause scumming, they should be analyzed daily to determine the amount of barium carbonate to be added. Frequent checks are necessary because soluble salts are migratory through leaching and evaporation, and locations of salt concentration can occur. These determinations assure enough additive to avoid scumming, and at the same time, prevent waste of expensive chemicals. The results of a turbidimeter test can be obtained in a relatively short time and the rate of addition can easily be corrected to meet the demand.

Raw materials purchased from outside sources should be checked in shipment lots for consistency with previous deliveries. The DTA can monitor uniformity of clay materials and reveal the concentration of troublesome impurities. Fusion buttons are appropriate for fluxes such as feldspathic materials. A partial chemical analysis may be the best quality determination for materials such as talc and pyrophyllite, where perhaps a SiO_2, MgO, or CaO measurement would be indicative of constancy. In a properly equipped laboratory, these analyses could be made in one day. Sometimes firing tests are necessary to ascertain the proper batch for color control.

The very important particle-size distribution usually needs to be completely determined only once a year or so unless the control tests indicate a significant change. An effective daily check on the distribution is a wet sieve analysis for the texture, filler, and plastic fractions.

In factories where grinding and screening are done, the structural clay products industry makes little or no attempt to maintain a constant moisture content previous to processing. Since increased dampness will alter the particle-size distribution towards a higher plastic fraction, frequent tests should be made on the raw materials entering the grinding process. These results can be used to modify the processes along the line in order to maintain uniformity. With finer grinds one would expect the water content for maximum plasticity to be greater, the drying time to be longer, and the firing temperature to be slightly lower to achieve uniformity in absorption.

The next critical point for a quality-control test is the water content at the time of forming. Because the amount of forming water is so important to the whole manufacturing operation, automatic control instruments at the pug mill are commonly employed to keep the water within narrow limits. In the absence of automatic controls, quality control checks should be made at least three times a day. Since immediate feedback is necessary, direct measurement of moisture content may not be possible unless an electrical resistivity measurement can be calibrated with respect to water content. The next best procedure is to measure plasticity by a quick penetrameter test or a direct measure of plastic strength. It must be remembered that plasticity can be affected by both particle-size distribution and water content; however, in most cases a check on plastic strength is sufficient, even though it is reduced by both insufficient and excess water. Usually it is obvious to the tester which side of optimum is prevailing.

In modern factories, conditions of drying and firing are automatically controlled and the control points are set by the superintendent. Data on conditions within the dryer and the kiln are recorded for future reference by those managing the quality-control program. In the absence of automatic controls, close manual control is required and conditions such as temperature, humidity, and firing atmosphere should be measured frequently to make sure they are within the desired limits at the time and place of measurement.

Quality-control personnel must monitor the final results of the entire operation by testing and examining the products on a daily basis. Production losses should be determined by a careful sampling technique. Size and absorption measurements must be made and plotted as shown in Fig. 116. Of course, these values must be derived from a proper sample representing the whole production. Color controls must be established by comparison with preset standards. Color variations are detected best with the human eye under standard lighting conditions, but they can also be recorded photographically. It is unnecessary to measure strength of face bricks, since if every other property is in its proper limits, there will be more strength than required; however, sewer pipes require test for strength because it is one of the primary specifications. Complete plans and procedures for developing a quality-control program for sewer pipes has been described by Ligon [2].

The recording of quality-control data should be tabulated in such a way as to allow direct tracing of a unit of raw materials through all the processes to the final product. This requires a system of keeping track of all materials moving through the plant.

9.4. Statistical Approach

Statistical quality-control procedures have not been used extensively in the structural clay products industry, but perhaps they should be mentioned here for their potential, not only for keeping production in line, but for a better understanding of the variables in the process. Perhaps the statistical analysis of an operation that contains so many interrelated variables has been too difficult and time-consuming. It is also true that many of the important results of a statistical analysis of the operation are learned by experience and close observation over an extended period of time. Be this as it may, a careful statistical analysis would produce useful and interesting relations that had not been clear before.

Thompson and Seal [3] have discussed the quality-control problems in a clay-working industry, and they have presented experimental designs which seem appropriate for the structural clay products industry. Some basic statistical procedures for use in the development of a quality-control program have been given by Hackler [4, 5], and text-books are available for background information [6].

References

1. Brownell, W. E.: Quality control in brick manufacturing. Am. Ceram. Soc. Bull. 41, 500–503 (1962); Claycraft 35, 419–23 (1962).
2. Ligon, E. R.: Process controls in the sewer pipe industry. Am. Ceram. Soc. Bull. 38, 269–73 (1959).
3. Thompson, H. R., and K. E. Seal: Serial designs for routine quality control and experimentation. Technometrics 6, 77–98 (1964).
4. Hackler, W. C.: Process Control, in Current Developments in the Whitewares Industry, G. A. Kirkendale, Ed., New York State College of Ceramics, Alfred University, June 1966.
5. Hackler, W. C.: Statistical Methods in Systems Design, in Systems Engineering in Ceramics, Nat. Bur. Stds. (U.S.) Misc. Pub. 267, pp. 109–15, 1965.
6. Miller, I., and J. E. Freund: Probability and Statistics for Engineers. Englewood, N.J.: Prentice Hall. 1965.

10. Plant Layout and Design

10.1. Predesign Planning

The capital investment for a new plant in the structural clay products industry is considerable. Because the unit price for the product is low, there is a minimum size for a plant and its daily output in order to be profitable. Modern brick and structural-tile factories are usually designed in production units of 100,000 standard brick-equivalents per day, and pipe plants should have a minimum output of 120 tons of ware per day. Probably the economic break-even point for face-brick production is about 60,000 units per day.

Careful planning should precede specific design and construction of a new facility. It must be established that there is a market for the quantity of products which must be made in order to be profitable. General consideration should be given to all elements of production of the products to be manufactured, such as availability of raw materials, plant location, process steps that will be in continuous operation and those which will be periodic, types of equipment to be employed, buildings, and the talents of the personnel that will be required. The expense of time and money in predesign planning is well spent because a bad plant is, so to speak, forever.

A competent and experienced engineering staff within the company considering new plant facilities is essential for a successful venture. It is possible and, indeed, highly recommended to use the knowledge provided by architects, by electrical, plumbing, and heating contractors, equipment suppliers, and consultants where necessary; however, the structural clay products company should evaluate all recommendations and make all final decisions on every aspect of planning, design, and construction. In the final analysis no one knows more about the whole process of manufacturing clay products than those doing it. The responsibility for a successful operation ultimately rests with the owners.

Consideration should be given to the type and number of employees required. An assessment should be made of the continuous cost of labor with its personnel problems, absenteeism, and turnover versus the capital investment in automatic machinery, its maintenance, and lack of versatility [1]. The processes where such decisions have special importance are mineral dressing, hacking, unloading, blending, and packaging. From the outset, the plans should contain provisions for plant cleanliness, because a clean plant will promote careful work from employees and reduce the maintenance costs of machinery.

General considerations and evaluations should be given to the proposed plant site, since the variables affecting the site are not always of equal importance. Cost evaluations should be made with regard to the proximity of the factory to markets, raw materials, energy supplies, and transportation. The precise location, once established from the concerns just mentioned, should be examined for its ability to hold secure foundations, for its potential danger by flooding, and for its liability to earthquakes, landslides, etc.

Before planning goes too far, the acquisition of raw materials must be considered. If raw materials are to be purchased from a supplier, will he be able to furnish them in the quantities required for the anticipated lifetime of the plant? If the raw materials are to be mined by the company operating the plant, the deposit should be evaluated as described in Chapt. 3, and the determinations of plasitc strength, drying behavior, and firing schedule should be made in time for use in the design of equipment.

Another question to be considered in the predesign stage is whether or not future expansion of production is to be designed into the plant. Since allowance for expansion in plant layout will add to the immediate capital investment, either expansion should be a good possibility within a 5-year period or it may be prudent to eliminate the allowance for expansion. Technology is developing so rapidly in these times that if expansion is estimated to be 10 years or more in the future, it may be better to build another plant at that time.

Just previous to active plant layout and design, tentative plans for the quality-control program should be made. The points for testing and evaluation can be established and the whole program integrated into the production system. A complete and detailed quality-control program will provide data invaluable in the start-up and shake-down periods.

10.2. Factors Affecting Plant Design

Designing a structural clay products plant really starts with decisions on the product to be manufactured and the production output desired. These factors determine the amount of material to be processed daily which, in turn, establishes the storage capacities at the various stages of manufacture. They also determine the size and numbers of units of equipment necessary to achieve easily the production scheduled. Like most process designs, it is good practice to make the various operational units a little larger than the minimum required, in order that the factory will not be constantly straining to meet full production.

Since there are many arrangements possible in plant layout, time should be allowed to study every angle of the designs first proposed. Starting plans should be tentative, to allow for changes if better arrangements are suggested. The whole process of establishing final plans may take up to a year or longer, and it has been the experience of the industry that such careful planning is time well spent. First, the layout must fit well into the geographical site, then the whole process must flow easily and economically without bottlenecks. Gravity should be used wherever possible to transfer materials, and thoughts should be given to each stage that requires power lifting of materials or products to higher levels to determine if such handling is really necessary.

The general shape of the plant layout determines the size and shape of the buildings which will house the operations. Since the buildings can be 10–14% of the plant costs [1] considerable thought should be given during layout planning to keep the costs of building construction from running excessively high. For example, a straight-line flow would require a building at least 1000 ft. (305 m) long and perhaps 100 ft. (30 m) wide. Such a building takes up a lot of real estate, is inconvenient for operating personnel and is probably a costly arrangement; therefore, a better plan would provide for a more compact arrangement. The situation to watch out for in going towards compaction is the length of roof spans, which rise rapidly in cost with length. Of course, columns to support roofing can be placed in convenient locations.

One problem which arises in making plants more compact is right-angle transfers. Since kiln cars run on rails, they are essentially adapted to a straight-line operation; so in breaking up the unidirectional flow into rectangular or parallel designs, power-operated transfer cars running on transfer tracks are necessary to convey kiln cars in right-angle directions to their tracks. Fig. 117 shows a transfer track to reverse the direction of kiln cars after they have been unloaded near the packaging station.

Fig. 117. Transfer track at the end of the production line

A close-up view of an automatic, electrically powered transfer car is given in Fig. 118. The problem here is that transfer cars are expensive; therefore, as few transfer arrangements as possible should be included in the plant layout.

In addition to the equipment and operational spaces in the building, one must remember to include space and facilities for offices, control instruments, quality-control laboratory, lavatories, lunch room, machine shop, electrical and electronic shops, maintenance shops for dies and molds, and kiln-car overhaul. All of these facilities should be conveniently located.

Fig. 118. Electrically powered automatic transfer car, designed to carry kiln cars from one track to another

As one follows through the process of making structural clay products from raw materials to finished products, there are a number of points of design that should not be overlooked in the layout. Raw material storage capacity should be large enough to avoid production shutdowns that might be caused by ordinary delays in mining, equipment failures, transportation tie-ups, or suppliers difficulties. Usually, the factories form ware eight hours a day, five days per week, except when holidays reduce the work week to four days. Since the kilns, and sometimes the dryers, operate on a continuous schedule, it is necessary to form enough ware during the working hours to keep the kiln supplied. This kind of scheduling requires sufficient storage tracks or areas for green ware for use when the forming process is not in operation. The plant layout should allow for quick transfer of ware from dryers to kilns to prevent excessive readsorption of moisture by the clay products. Kiln cars require continuous superficial maintenance and periodic overhaul; therefore, the layout should provide a cleaning and inspection station in the regular flow pattern and a diversion track to lead cars into the repair shop. In normal operation, it is usually necessary to have at least one car in the shop at all times. Some face brick factories require a blending line for color mixtures after the ware has been fired, and the blending operation should be conveniently located to packaging. Packaged inventory can be in covered storage or in an open paved yard, whichever seems to be most appropriate. Throughout the whole layout dust, air, noise, and water pollution controls must not be forgotten in the design of a new plant.

A parallel activity during the planning and construction of a plant is the acquisition of personnel who will be responsible for operation and maintenance. During construction, most pipes, wires, and internal parts of equipment are exposed. Growing up with the plant is an invaluable training program which will pay handsome dividends in starting up and in smooth operation thereafter.

10.3. Planning for Starting the Plant

Starting up a new plant should be planned carefully because 5–6% of the plant cost is involved here [1], and the cost could be much higher if a good program is not established. In such a complex system of interdependent operations, troubles and adjustments are inevitable. How quickly they are analyzed and corrected will make a great difference in the time to normal operation when profits begin.

Several prestarting procedures are highly recommended. Personnel should be pretrained on the plant and its equipment by observing during construction and by studying equipment manuals, piping diagrams and electrical drawings. Exercises in the proper procedures to follow in case of specific troubles are excellent training procedures. A documented maintenance program should be laid out for all units of equipment. Log books should be organized to indicate the proper maintenance procedures, and space should be provided for dates and initials to show that the work was done. Before starting up a plant, job descriptions should be written out and given to all employees so that everyone knows his responsibilities. The act of preparing such job descriptions assures the management that all of the necessary functions are included. This procedure leads to smooth operation, and much periodic confusion is avoided. Finally, just before starting the intraplant communication system should be checked for complete and proper functioning.

References

1. Brandt, W. O.: Planning for a New Plant. Am. Ceram. Soc. Bull. 50, 185–88 (1971).

11. Serviceability and Durability

11.1. Serviceability of Structural Clay Products

With structural clay products, the proof of the pudding is their ability to resist all the rigors of service conditions and to function well in their intended purposes for very long times. When one builds his home, factory, or place of business using clay products, he likes to think that it will last forever with no maintenance, and perhaps it might, if it did not become obsolete, outgrown, or exposed to natural catastrophes. Furthermore, the durability of structural clay products is expected to exceed other materials such as wood, metals, and plastics, and indeed it will, as evidenced by the antique products up to 10,000 years old illustrated in Chapt. 1. Bricks and tiles are expected to produce strong buildings and pavements. Roofing tiles must be designed to keep out rain and snow, and pipes for the transportation of all kinds of wastes are designed to produce leak-proof joints with a minimum of skill and labor.

In addition to durability, face bricks are involved with esthetics. They must be decorative as well as serviceable. The wide variety of colors and textures available today allow architects to express the purpose of the building in their design. For example, white or clean-looking bricks are often used for hospitals, food-preparation plants, research facilities, and utility buildings where cleanliness is of prime importance in successful operations. Colors and textures are often selected to blend into the landscape so that the building seems to belong. When bricks are selected for such applications, they are expected to remain attractive for the life of the facility no matter how long that may be.

Floor and wall tiles are expected to resist staining and to never require replacement or resurfacing in use. These products are often subjected to excessive dampness, and they must be cleanable with antiseptic and germicidal chemicals as in bathrooms, public lavoratories, operating rooms, swimming pools, and food processing areas. Tiles are also expected to withstand whatever abrasion and ware are encountered in normal use.

Sewer pipes must withstand the pressures of overlying earth, and the stresses of possible shifting and subsidence of the bed on which they are laid. Once a sewer line is installed, no one expects that it will have to be dug up for repair or replacement.

11.2. Durability of Bricks and Roofing Tiles

The durability of bricks and roofing tiles refers to their resistance to the agents of weathering—principally rainfall and freezing temperatures, but sometimes also exposure to soluble salts such as sodium chloride from sea water, and sulfates from industrial air pollution.

The location and climate where the products are used determine the severity of weathering. In hot, arid regions the only requirement is strength enough to endure the loads and stresses of the building itself. In fact, adobe (unfired clay) bricks are quite serviceable under these conditions. In more temperate climates the elements of precipitation and freezing temperatures demand very special properties in structural clay products. Cyclical freezing and thawing temperatures are particularly destructive weathering forces on wet products. Ironically, many of the highly populated areas of the world are in severe-weathering zones, where people need the protection and comfort provided by bricks and roofing tiles.

The characteristics of clay products that determine their durability on exposure to moisture and freezing temperatures are the phases present in the fired product, elasticity, strength, and porosity. Of these, porosity seems to be the predominant factor because, in most cases, when the pore structure is right for resistance to freezing cycles, stable phases and sufficient strength will also have been developed by proper firing.

The types and amounts of porosity in a clay body with satisfactory resistance to frost action are quite complex. The key seems to be related to the number of pores and how they are connected by finer capillaries. Across all types of fired clay materials, total porosity does not appear to be a factor, but in any specific material there is a relation between total open porosity and resistance to the freezing of pore water.

In the industry, open porosity is generally measured by water absorption, and in some laboratories mercury intrusion is used. The advantage of the former is simplicity and that of the latter is that pore-size distribution is determined. It is common practice to measure the water absorption of bricks after soaking 24 hours in cool water and after boiling in water for 5 hours; therefore, there are two values for water absorption.

By strictly empirical methods, the U.S. National Bureau of Standards found a reasonable correlation between a saturation coefficient and the ability of a brick to withstand freezing and thawing cycles while in contact with water [1, 2, 3, 4]. This *saturation coefficient (C/B ratio)* was derived by dividing the percent absorption after the 24-hour cool soak by the percent absorption obtained by the 5-hour boil. As a result of this work, ASTM Committee C-62 specified a saturation coefficient for bricks. For bricks that might be exposed to severe weather conditions, C-62 specified that the saturation coefficients for 5 bricks should average not more than 0.78 and no single brick should have a saturation coefficient exceeding 0.80. Of all the specifications for bricks, this is the most important one for resistance to frost action. Over the years, a low saturation coefficient has been found to be a reasonable measure of durability except when artificially induced by the inclusion of a burnout material [5] and when the total absorption is below 2%. For any particular clay material, the C/B ratio decreases from values near 1.00 for low firing temperatures, to values approaching 0.50 for high temperatures. Thus the saturation coefficient is indicative of the firing maturity of a clay product.

During the firing of clay products the total pore volume decreases with increasing temperature, and of course this is accompanied by shrinkage. In addition there is a

change in the pore-size distribution. Two pore-size distributions for a typical structural clay material prepared by vacuum extrusion and fired to different temperatures,
are shown on Fig. 119. The pore-size data were obtained by mercury intrusion. The
1800°F (982°C) firing produced a product nonresistant to frost action, but the
1950°F (1066°C) specimens were resistant to more than 50 cycles of freezing and
thawing. The plots on Fig. 119 used the phi transformation discussed in Chapt. 3 for
particle-size distributions. The solid line gives the pore-size distribution for an underfired situation where the C/B ratio was 0.85, and the total boiled absorption was
about 15%. The dashed line is for the same material properly matured by firing to
1950°F, which resulted in a C/B ratio of 0.66 and a total boiled absorption of about
5.5%. From the two distributions one can see that most of the pores were in the one-
micrometer range, and that proper firing reduces the total pore volume, shifts the
median to slightly larger pores, and greatly reduces the volume of pores below the
0.6 micrometer range.

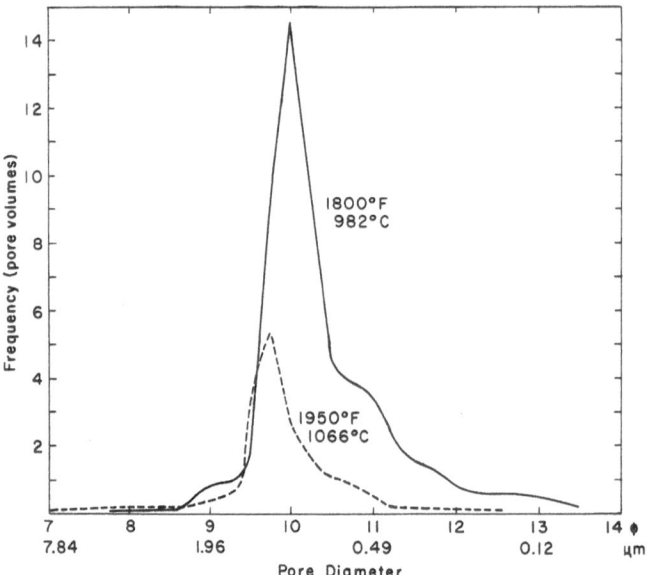

Fig. 119. Pore-size distributions for a clay body fired to two temperatures—underfired and normally fired

What is the physical meaning of the C/B ratio or saturation coefficient? It is a
kind of measure of pore structure. A high saturation coefficient indicates that most
of the open pores are easily filled by submergence in water, while a low C/B ratio
means that simple soaking will not fill all of the open pores—perhaps because of a
structure that allows the entrapment of air. Values approximating the 5-hour boiled
absorption can be obtained by applying a vacuum over the specimen submerged in
cold water. The saturation coefficient is then written C/V. As previously indicated,
the saturation coefficient is unrelated to total absorption when one considers all
kinds of basic raw materials from which structural clay products are made. Coefficients below 0.80 can be achieved with products having total absorptions between

about 4% and 18%. Proper firing tends to close the accessibility of water to pores, probably by breaking up the continuity of the pore structure through eliminating many fine capillaries and reducing the total number of pores. If the pores are not filled by exposure to water, there will be space for the ice to expand without placing large stresses on the product's internal structure.

The freezing of water in a clay body has been observed to occur in two stages [6]. First, after some supercooling, water freezes in larger pores at about 0°C, accompanied by some expansion. As the water freezes in the larger pores, additional water tends to flow from the connecting capillaries into the pores where it freezes. This migration continues until the pressure on the ice phase raises the chemical potential of the ice in the pores until it equals the chemical potential of capillary ice. In the case of clay products only a little water can flow from capillaries to pores, because the rigid pore walls force the pressure up quickly—unless, of course, fractures (microcracks) occur in the pore walls. Second, water begins to freeze in the capillaries as the temperature is lowered, and more expansion takes place. The silicate-surface structuring of the water in the capillaries increases the chemical potential of capillary ice; therefore, the freezing of water in capillaries occurs progressively in finer and finer capillaries as the temperature is lowered. In clay products the capillary water freezes between about $-2°C$ and $-8°C$. Generally, more expansion takes place during capillary water freezing than was generated by the freezing of water in the larger pores.

The pressure exerted by water freezing in clay products, according to the capillary theory of frost damage when it is assumed that the ice and water phases are in equilibrium with the saturated vapor outside the sample, is expressed by [6]

$$P_c = \frac{\Delta S_f (T_f - T_0)}{V_f} \tag{1}$$

where
P_c is the ice pressure,
ΔS_f is the entropy of fusion,
T_f is the temperature of freezing in the finest capillaries,
T_0 is the melting temperature of bulk ice at atmospheric pressure,
V_f is the molar volume of ice at T_f.

The capillary theory of frost pressure as expressed in Eq. (1) represents the minimum pressure exerted on a ceramic body. It does not allow for localized ice pressures in excess of those required for equilibrium. Such localized high pressures have been observed in ceramic bodies during freezing, because of the broad range of pore sizes, capillary sizes, and types of connections among them [6].

When the ice phase sustains pressure gradients under nonequilibrium conditions between pore ice and capillary ice as, for example, if water becomes trapped in capillaries by pore ice, the localized pressures could equal that predicted by the ice-water phase diagram based on the Clapeyron-Clausius equation [6]

$$\ln P = -\frac{\Delta H_f}{R}\left(\frac{1}{T_f} - \frac{1}{T_0}\right) \tag{2}$$

where
P is the pressure on the ice phase,

T_f is the temperature of freezing in the finest capillaries,

T_0 is the melting temperature of bulk ice at atmospheric pressure,

ΔH_f is the heat of fusion of water to ice,

R is the gas constant.

Eq. (2) predicts pressures an order of magnitude greater than the capillary ice theory, but since some relief of pressure can be expected to occur, it represents only an upper limit to the pressure generated by freezing water in clay bodies.

Another factor besides pore volume and pore structure which determines whether or not frost damage will occur is the elastic strength of the internal structure of the clay product. Well-fired clay products have a better chance of resisting frost damage because of their increased modulus of elasticity, developed by proper phase development. In addition to increased strength, well-fired products also have reduced pore volume and fewer connecting pores and capillaries, both of which favor resistance to frost damage.

These properties have been experimentally evaluated for commercial bricks known to have both high and low resistance to frost damage [6]. The data are listed in Table 15

Table 15. *Factors Affecting Resistance to Freezing and Thawing of Water in Bricks* [6]

	High Resistance	Low Resistance
Total pore volume (%)	16.9	23.3
Range of pore diameters (μm)	0.28–23.0	0.04–34.0
Average pore diameter (μm)	1.60	0.24
Young's modulus of elasticity (psi)	3.6×10^6	1.5×10^6
C/V saturation coefficient	0.72	0.90
P_i (psi) – actual ice pressure	1680	3400
P_c (psi) – capillary theory ice pressure	< 570	< 1200

where it can be seen that the more resistant brick had a lower pore volume, a narrower pore-size distribution, a larger average pore diameter, and a higher modulus of elasticity. Notice also that C/V is below the specified 0.80 for the resistant brick and above this value for the one with low frost resistance.

In this table pressure P_i was calculated from the measured freezing expansion and the modulus of elasticity of the specimen. P_c was calculated from Eq. (1). Note that the actual pressure exerted by the ice on the internal structure was considerably higher than that predicted by the capillary theory of frost damage.

When a clay product has a low resistance to the freezing of water in its pores, each freezing-thawing cycle adds microcracks to those inherently present. Usually several cycles are necessary for complete and visible failure due to the ever increasing number of cracks developed with each additional cycle. ASTM C-62 presumes a brick to be resistant to frost action if it withstands 50 cycles of freezing and thawing under saturated conditions in laboratory testing.

For maximum resistance to the action of moisture and freezing temperatures, there is no substitute for a well-fired product. This statement applies equally to bricks, roofing tiles [7], terra cotta, sculpture, or pottery that might be exposed to the weather. Glazed products are particularly sensitive to this type of deterioration if water finds its way into the body of the product.

11.3. Durability of Sewer Pipes

The different environmental conditions and the different service requirements of sewer pipes direct attention towards different properties as a measure of durability than those described for bricks and tiles. To insure against seepage ASTM Committee C-13 specifies that the 5-hour boiled absorption of sewer pipe bodies should not exceed 8%. Since the service conditions of buried pipes subject them to external pressure, C-13 sets minimum crushing strengths in pounds per linear foot. The specified strengths vary from 1800 to 6000 lb./ft., as determined by a sand-bearing method, for small to large diameter pipes. Resistance to solubility in acids is also specified by C-13. Strength and acid resistance is promoted in sewer pipes by the formation of mullite crystals during the firing operation. For this reason kaolinitic or relatively pure illitic clays are usually used for the manufacture of these products.

11.4. Moisture Expansion

Structural clay products are usually thought of as the ultimate in dimensional stability; however, some of these materials are subject to expansion over very long periods of time, due simply to exposure to atmospheric moisture [8]. The magnitude of linear moisture expansion, for all kinds of clay products exposed to moisture over a two-year period, are from 0.004% to 0.186%. If an expansion of 0.1% were to take place in a brick wall 100 ft. (30.5 m) long, the wall would expand 1.2 in. (3 cm). Such a change in length could produce extremely serious structural damage. Under conditions of normal exposure, moisture expansion increases rapidly during the first 6 months and then slows exponentially with time, but some products may show some expansion after a 50-year period [9].

Moisture expansion is caused by the chemical adsorption of water vapor on the internal amorphous silicate surfaces that may be present in a clay product. It should be clear that the physical adsorption of water, which can be removed by a vacuum or heating to $110°C$, does not play a role in moisture expansion. Chemically adsorbed water can only be removed by heating to $705°C$ ($1300°F$) or higher [9, 10, 11, 12]. Incidentally, most, but not all, of the expansion is recovered in the form of shrinkage upon such heat treatment. Chemical reactions between the silicates and water to form crystalline hydrates during moisture expansion have not been established, but the adsorbed water could possibly produce layers of amorphous hydrates on all non-crystalline silicate surfaces. It has also been suggested that the expansion is caused by a relaxation phenomenon, when the high surface energies of amorphous constituents are lowered by adsorption of water molecules. The phases under consideration here are amorphous or very poorly crystalline silicates such as occur during the conversion of disilicate minerals to high-temperature crystals and glassy phases [9, 11, 12].

The production in clay products of amorphous silicates and glass which are responsible for moisture expansion, is related to mineralogy and composition. Na_2O and K_2O in the composition cause, each in its own way, amorphous phases to develop [10, 13]. Sodium oxide causes fusions early in the firing process, which result in rather unstable glasses, especially at first. Such fusions result in a small amount of a glassy phase dispersed throughout the body in thin films and fibers; as such, they generate extensive surface areas for moisture adsorption. At the same time, the fluxing action of sodium causes a consolidation of the body, resulting in fewer pores and capillaries for access of water vapor to these surfaces. Because of these opposing effects, the rate of moisture expansion is reduced by sodium oxide at short times, but it continues at appreciable rates for a very long time—perhaps 50 years or more [14, 15, 16]. Potassium oxide first contributes to moisture expansion by retarding the crystallization of aluminosilicates, preserving amorphous phases to higher temperatures; later it produces fusions that result in glassy phases in the body [15].

Porous glass specimens have been observed to have very high moisture expansions, well in excess of those encountered in structural clay products. They expand even when nonpolar molecules are adsorbed [17].

The discussion of silicate reactions with alkaline earth oxides in Chapt. 6 will suggest compositions to reduce or practically eliminate moisture expansion. Magnesium, calcium, strontium, and barium oxides all act to reduce the amount of amorphous aluminosilicates by rapid reactions to form well-developed crystals at low temperatures [13, 14, 15, 16]. These reactions also tend to prevent the formation of liquid phases that might produce a glassy bond on cooling. In structural clay products, firing temperatures are rarely high enough to cause fusion of the alkaline earth silicates. In line with these suggestions, naturally occurring high-lime clays and shales exhibit negligible moisture expansions when properly prepared and fired.

The smallest of the alkali ions, lithium, produces an effect opposite to the rest of this family. When lithium carbonate is added to clays, it tends to form crystalline lithium aluminum silicates before melting occurs; therefore, lithium acts to reduce moisture expansion in clay bodies in a manner similar to that of the alkaline earth oxides [15].

It is certain that pore structure and volume affect moisture expansion, since high porosity allows easy penetration of water molecules to the interior surfaces of the body; however, data from tests on commercial bricks demonstrate the dominance of composition over porosity in tendency toward moisture expansion. An illitic clay brick, high in K_2O and having an absorption of 10.5%, produced 0.195% linear moisture expansion after treatment in an autoclave. Companion experiments on a high-lime, commercial brick with an absorption of 16.1% showed a moisture expansion of only 0.008% [14].

Since moisture expansion is a long-range process, products cannot be adequately tested before use; therefore, there has been a tendency to use an autoclave to accelerate moisture expansion. Attempts have been made to correlate time, temperature, and pressure of autoclaving to years of normal service. To a certain extent this may be possible, but it is very likely that the increased temperature of the autoclave promotes hydrations that would never occur under atmospheric conditions, because of the higher activation energies that may be reached. Moisture expansion can be observed quickly with autoclave treatment, but care must be taken if the results are to be meaningful in predicting normal behavior [11].

In summary, the rate of moisture expansion depends on porosity, while the extent depends on the chemical and mineralogical composition of the fired product. Where moisture expansion becomes a serious concern, composition adjustments can be made to reduce the problem to a negligible level. As in so many cases of service problems, there is no substitute for a well-fired product. Proper firing favors stable crystals, more chemically resistant glassy phases, and increased crystal perfection and size, all of which tend to minimize moisture expansion. In some construction projects, the use of expansion joints, soft lime mortar, and brick aging are good practices to avoid moisture expansion problems.

Moisture expansion becomes a most critical factor when any clay product is glazed. Even when a well-fitted glaze starts off under compression stress, the gradual expansion of the body in use may overcome the compressive forces and eventually place the glaze in tension. When this happens—perhaps a year or more after installation—crazing of the glaze occurs. If the glaze substrate is not particularly strong, the crazing cracks will extend into the body, and spalling is apt to occur. Products especially susceptible to this action are glazed bricks, structural tiles, floor and wall tiles, architectural terra cotta, and roofing tiles. With these products attention must be paid to composition as well as heat treatment. ASTM specification C-126 for glazed products calls for no crazing after one hour in an autoclave at 150 psi.

The action of moisture expansion is to put internal stresses in the body of clay products, and some of the localized stresses are relieved by microcracking. Such relief of stresses is demonstrated by the fact that only part of the moisture expansion of clay products is reversible. Since all structural clay products inherently have a certain concentration of microcracks, moisture expansion gradually adds to this number, further weakening the body. It has been found that the strengths of clay products are often less after autoclaving, and the products are also less resistant to freezing and thawing [18].

In poorly manufactured clay products, the progressive increase in the number of microcracks eventually leads to macrocracks and complete failure after some time in service. Of course, this behavior is to be avoided at all cost.

11.5. Bonding of Mortar to Bricks and Tiles

The quality of the bonds between structural clay products and mortar is primarily determined by construction skills over which manufacturers have little control, except in the case of prefabricated panels; however, porosity tends to determine bonding effectiveness with otherwise good mortar and workmanship. It is, then, prudent for those making structural clay products to provide optimum bonding properties in their wares.

The property of bricks and tiles most related to mortar bonding is the *Initial Rate of Absorption (IRA)*. ASTM prescribes an IRA test, in Specification C-67, wherein a dry brick is immersed flatwise in 1/8 in. of water for one minute. IRA is expressed as the grams of water absorbed by 30 square inches of surface area. Experience has shown that IRA values from 10 to 20 grams give the maximum number of good bonds in a brick wall [19]. Bricks with lower IRA may theoretically give tight bonds, but the rate of bond development between such bricks and mortar is so slow that bricklayers have difficulty in maintaining a coherent initial set while the work progresses. Bricks with higher IRA tend to absorb the water from the mortar paste before the initial set takes place; thereby, weak bonds are produced, because the mortar next to the bricks does not have enough moisture for proper hydration and setting. Bond improvement is best achieved with high IRA bricks by presoaking, but the brick manufacturer has no control over such procedures on the job site. The next best procedure is for the manufacturer to dip his products in a silaneal solution to lower the IRA to about 20 grams [20]. Overall this is probably the better practice because the bricks are sent to the job with absorptions conducive to good bonding, without further preparation previous to laying in a wall.

There must be a relation between IRA and total absorption for clay products, but the relation is not simple and is affected by pore structure and capillary rise. Different methods of manufacture, such as soft-mud forming, stiff-mud extrusion, and dry pressing, produce different IRA for products having the same total absorption. In addition, variations in IRA for products made by the same method occur when raw materials having different particle-size distributions are used.

Experiments have shown that thin glazed wall tiles and unglazed floor tiles with absorptions between 2.4% and 12.9% give the best bonds with cement mortars if they are not prewetted or presoaked [21].

11.6. Efflorescence and Staining of Brickwork

Efflorescence is a salt deposit formed on the surface of clay products when they have been in contact with water [22]. Soluble salts can adversely affect the durability and appearance of brick buildings and roofs made of tiles [23], but efflorescence is most commonly associated with disfigurement of brickwork by staining. The mechanisms of salt deposition are many and complicated, but simply, soluble salts are brought to the surface of brickwork by solution in water and deposited there by evaporation. The salt solutions can migrate across the faces or through the pore structures of bricks and tiles. The sources of soluble salts may be the product itself, mortar cements, construction backup materials, sulfur gases in the atmosphere, or sea water and mists [22].

The most common salts to form on uncontaminated clay products are sulfates of calcium, magnesium, sodium, and potassium. Of all these, *calcium sulfate* is the most troublesome because it is the most common effloresing salt and the most difficult to remove. Calcium sulfate is a common impurity in clay raw materials, and it can persist through the firing operation, since the temperatures are often not high enough to decompose the crystal or cause it to react completely with the silicates. (These high-temperature reactions were discussed in Chapt. 6.) The other salts mentioned above cannot resist the firing operations without melting, decomposing, or reacting with silicates, but they can still appear as effloresing salts on products tested directly from the kiln. This behavior is possible when gaseous SO_3 is adsorbed on the internal silicate surfaces of the ware by exposure to sulfurous gases during firing and cooling. In this case sulfuric acid is formed as water is drawn into the product by capillary action, and this corrosive acid will dissolve magnesium, sodium, and potassium from various crystalline and glassy phases. When these solutions migrate to the surfaces, evaporation concentrates them to the point of saturation, and salt deposition occurs. Another effect of sulfurous-gas contamination during processing takes place when calcium carbonate is present in the raw materials. In this case the carbonate is converted easily, even in the dryer, to calcium sulfate where it then behaves as if it were originally associated with the clay [22].

Vanadium salts also cause troublesome efflorescence in the form of green vanadyl sulfate $[(VO)_2(SO_4)_3]$ and brown vanadic acid (H_5VO_5). As might be expected, deposits of these compounds are particularly unsightly on light-colored products, and as nature would have it, the raw materials for white or buff bricks contain the most vanadium. Clay minerals, especially kaolinite, can contain a few tenths of a percent of vanadium, present as a trivalent substitional solid solution for aluminum. On firing, the vanadium is released from the clay structure and oxidizes to the pentavalent state. The vanadium pentoxide has a low melting point but does not react with either silica or alumina; however, it is soluble in water. As water leaches through the pore structure of the bricks, it dissolves some of the vanadium pentoxide, and if sulfates are also present, vanadyl sulfate precipitates on the surfaces. In the absence of sulfates, brown vanadic acid precipitates. At times, the cleaning of brickwork with muriatic acid will bring out blue-green vanadyl chloride $(VOCl_2)$ crystals [22].

In addition to the soluble salts originating within products, salt solutions may migrate into the pore structures from outside sources and cause efflorescence during periods of surface drying. If the source of soluble salts is continuous over several months or longer, severe damage to the products in the form of flaking and spalling can be the result of salt crystallization in the pores. The effloresing salts from direct outside contamination are sodium chloride from the sea, calcium hydroxide and carbonate from concrete backup materials, and alkali carbonates from mortar. The latter are often observed as *"new-building bloom,"* and they result from the free-alkali content of the cement used in the mortar. These salts appear only once, shortly after construction, since after the mortar has achieved a final set, the joints become relatively impervious to further leaching. If the new building is allowed to dry thoroughly before cleaning, these alkali carbonates are easily removed by washing, and they will not reappear [22].

Another type of efflorescence can occur on masonry work due to excessive cleaning with muriatic acid and failure to rinse throughly with water. The acid readily attacks the mortar joints creating a solution of potassium chloride. When KCl appears as the principal salt of efflorescence, the wall should be allowed to dry thoroughly, scrubbed with a dry brush, and finally rinsed with pure water [22].

Sometimes the chemical incompatibility between bricks and mortar results in *alkali sulfates* as efflorescing salts. The adsorbed sulfate molecules on the internal surfaces of some bricks cause acidic solutions to form when water enters their pore structures. This acid then attacks the mortar joints, releasing sodium and potassium ions into the solution. Heavy efflorescence of this type can occur only after the bricks have been exposed to a sulfurous atmosphere on cooling or in inventory. Often such bricks will show no efflorescence in laboratory tests because the phases present are resistant to sulfate attack [22].

The *manganese staining* of mortar joints is somewhat related to the efflorescence problem. When manganese dioxide is added as a body stain to red-firing products to produce shades of brown and black, a brown stain sometimes appears on the mortar joints a few weeks after construction. The mechanism of development of this stain is set forth by

$$4MnO_2 \rightarrow 2Mn_2O_3 + O_2 \tag{3}$$

$$2Mn_2O_3 + 4H_2SO_4 \rightarrow 4MnSO_4 + 4H_2O + O_2 \tag{4}$$

$$MnSO_4 + 2KOH \rightarrow Mn(OH)_2\downarrow + K_2SO_4 \tag{5}$$

$$\underset{\text{white}}{6Mn(OH)_2} + O_2 \rightarrow \underset{\text{brown}}{2Mn_3O_4} + 6H_2O \tag{6}$$

Manganese dioxide decomposes according to Eq. (3), on firing to $1090°F$ ($588°C$); thereafter, the manganese oxides resemble the oxides of iron in their high-temperature reactions. All three high-temperature oxides of manganese (Mn_2O_3, Mn_3O_4, MnO) are relatively soluble in acidic solutions, and their solubility increases in the order given. In commercial practice, reducing conditions are often used during the firing operation to intensify the dark colors of manganese-containing bricks, thereby increasing the solubility of the manganese. Eq. (4) was written for the solubility of Mn_2O_3 in a sulfuric acid solution derived from adsorbed SO_3 molecules on the internal silicate surfaces; however, it could just as well have been written for any of the manganese oxides, or using hydrochloric acid derived from the brickwork cleaning operation. The resulting acidic solution of manganese diffuses across all exposed surfaces of the wall, including the mortar joints. On contact with the highly basic mortar containing potassium hydroxide (KOH), neutralization and precipitation of manganese hydroxide takes place immediately, as expressed by Eq. (5). At first the hydroxide is an invisible white deposit, but on exposure to air and drying, the hydroxide changes to the brown manganese oxide, Mn_3O_4. This is the mechanism and nature of the brown staining of mortar joints [22, 24, 25].

Manganese dioxide has been used for many years to darken fireclay products without trouble from brown staining. The reason is that the higher firing temperatures promote a reaction, especially under reducing conditions, to form the manganese aluminum silicate, spessartite $[Mn_3Al_2(SiO_4)_3]$. This crystalline phase is insoluble in acids; therefore, the brown-stain mechanism cannot get started. The identification of spessartite in nonstaining products led to a suggestion that this silicate compound be prepared and added to low-temperature, red-firing bodies to achieve the color with a stable pigment. Spessartite is easily prepared by mixing and compacting kaolin and pyrolusite (MnO_2) and heating to about 2020°F (1104°C) [26].

11.7. Cleaning Brickwork

Although the cleaning of brickwork is not the responsibility of the brickmaker, he often gives recommendations on cleaning problems to prevent irreparable damage and to ensure customer satisfaction. In all cleaning operations it must be realized that the cleaning agents will enter the pores of the bricks, where they may produce deleterious effects for a long time, and may cause efflorescence or staining as the brickwork dries. Before any attempt is made to clean bricks, the undesirable deposits or stains must be correctly analyzed first, in order to insure the use of the proper cleaning agents. In many cases, presumptions have led to worse problems after cleaning. A factor often overlooked in cleaning operations is the chemical instability of the mortar joints. They are very sensitive to the action of chemicals, and the products of these reactions may continue to stain the brickwork.

New buildings must be cleaned soon after construction to remove mortar smears that result from normal masonry procedures. Contractors and customers should be patient concerning this first cleaning operation to allow the structure to dry thoroughly and the mortar to set completely. A delay in cleaning will allow the "new building bloom" to be removed along with the mortar stains, and chemical attack of the mortar joints will be minimized. A 10% hydrochloric acid solution is recommended for initial cleaning to remove the mortar smears. The Structural Clay Products Research Foundation published the following recommendations for cleaning new buildings [27]:

1. The masonry should be saturated with water as much as possible before the acid is applied.

2. Mortar smears should be removed by brushing with a 10% muriatic acid solution. The highest grade of acid should constitute the base of the solution.

3. Immediately after a small section of the wall has been cleaned, it should be washed thoroughly with water, as from a hose.

4. In cases where the building has been constructed with white, light buff, or manganese bricks, the brickwork should be acid neutralized, while still wet, with a 0.1 M solution of potassium hydroxide or carbonate. No rinsing should follow this neutralization treatment [22].

Before attempting to remove efflorescences and stains from brick walls, the offending materials must be identified, and their physical-chemical properties, as well as their mechanism of formation, must be taken into account. The proper cleaning agents and procedures will depend on these factors. Highly soluble salts should be brushed off, and the walls may then be rinsed with pure water if necessary. Calcium sulfate is only slightly soluble in water or acid solutions, and may be nearly impossible to remove completely. Many cases are best left to nature to remove by periodic rainfall. Vanadium stains must be removed with solutions of caustic soda (NaOH) or potassium hydroxide. Manganese stains on mortar joints can be easily removed by a mixture of acetic acid and hydrogen peroxide, but this must be followed by an acid neutralization treatment with a strong base [22, 27, 28, 29]. The detailed procedures for these cleaning operations are given in the references cited.

Chemicals and procedures for the removal of iron rust stains from brickwork have been published [27, 29]. A strong solution of oxalic acid, possibly with ammonium bifluoride added, can be used effectively. Another procedure prescribes glycerine-sodium citrate and sodium hydrosulphite-talc paste poultices.

Paint can best be removed from bricks and tiles while still relatively fresh—perhaps within a few days after application. Commercial paint removers or a solution of trisodium phosphate should be used first [27, 29]. A product containing isopropanol and ethanol has proven to be particularly effective in removing paint smears and the residue left by other paint removers. Some brick textures make complete paint removal almost impossible, but good success has been achieved with smooth, sanded, or lightly textured surfaces.

References

1. McBurney, J. W., and A. R. Eberle: Freezing and thawing tests for building brick. Proc. Am. Soc. Testing Mats. 38, Part II, 470–83 (1938).
2. McBurney, J. W., and J. C. Richmond: Strength, Absorption, and Resistance to Laboratory Freezing and Thawing of Building Brick Produced in the United States. Natl. Bur. Stds. (U.S.), Materials and Structures Rept. No. BMS60 (1940).
3. McBurney, J. W., J. C. Richmond, and M. A. Copeland: Relations among certain specification properties of building brick and effects of differences in raw materials and methods of forming. J. Am. Ceram. Soc. 35, 309–18 (1952).
4. McBurney, J. W., and P. V. Johnson: Durability of deaired brick. J. Am. Ceram. Soc. 39, 159–68 (1956).
5. Carlsson, O.: The Influence of Submicroscopic Pores on the Resistance of Bricks Towards Frost, Trans. Chalmers Univ. Tech., No. 212, Gothenburg, Sweden, 1959.
6. Blachère, J. R., and J. E. Young: Failure of capillary theory of frost damage as applied to ceramics. J. Am. Ceram. Soc. 57, 212–16 (1974).
7. Cole, W. F.: Terra cotta roofing tile deterioration in Australia. Brit. Clayworker 70, 126–33 (1961).
8. McBurney, J. W.: Masonry cracking and damage caused by moisture expansion of structural clay tile. Proc. Am. Soc. Testing Mats. 54, 1219–41 (1954).
9. Hosking, J. S., and H. V. Huber: Moisture expansion of clay products with special reference to bricks. Trans. 7th Int. Ceram. Cong., London, pp. 311–25, 1960.
10. Demeduck, T., and W. F. Cole: Contribution to the Study of Moisture Expansion in Ceramic Materials. J. Am. Ceram. Soc. 43, 359–67 (1960).
11. Young, J. E.: The Influence of Physical Adsorption of Water Vapor on the Moisture Expansion of Fired Kaolin, Ph.D. Thesis, New York State College of Ceramics, Alfred University, Sept. 1961.
12. Smith, A. N.: Investigations on the moisture expansion of porous ceramic bodies. Trans. Brit. Ceram. Soc. 54, 300–18 (1955).

13. Milne, A. A.: Expansion of fired kaolin when autoclaved and the effect of additives. Trans. Brit. Ceram. Soc. **57**, 148–60 (1958).
14. Young, J. E., and W. E. Brownell: Moisture expansion of clay products. J. Am. Ceram. Soc. **42**, 571–81 (1959).
15. Brownell, W. E., and J. E. Young: Effects of Composition on Moisture Expansion of Clay Products, New York State College of Ceramics, Mon. Prog. Rept. No. 269, 1961.
16. Cole, W. F.: Some relationships between mineralogical and chemical composition and moisture expansion of fired clay bodies. J. Austral. Ceram. Soc. **4**, 5–9 (1968).
17. Yates, D. J. C.: The expansion of porous silica glass produced by the adsorption of non-polar gases at liquid air temperatures. Trans. Brit. Ceram. Soc. **54**, 272–99 (1955).
18. Crupain, D.: Interrelation of Durability Tests on Glazed Brick, B.S. Thesis, New York State College of Ceramics, Alfred University, June 1961.
19. Ritchie, T., and J. J. Davison: Factors Affecting Bond Strength and Resistance to Moisture Penetration of Brick Masonry, ASTM Spec. Tech. Pub. No. 320, pp. 16–20, 1962; NRC (Canada), Div. Bldg. Res. Paper No. 192, Ottawa, 1963.
20. Carlson, B. C., and W. D. Betts: How plant-applied silicones affect wall strengths. Brick Clay Rec. **135**, 49–51; 68 (1959).
21. Waters, E. H.: The Effect of the Moisture Content of Ceramic Tiles on the Strength of the Tile/Mortar Bond, CSIRO, Div. Bldg. Res. Tech. Paper No. 7, Melbourne, Australia, 1959.
22. Brownell, W. E.: The Causes and Control of Efflorescence on Brickwork, Structural Clay Products Institute (now Brick Institute of America) Res. Report No. 15, McLean, Va., Aug. 1969.
23. Cole, W. F.: Some aspects of the weathering of terra cotta roofing tiles. Austral. J. Appl. Sci. **10** (3) 346–63 (1959).
24. Bebbington, P. J., and R. C. Feagin: Discoloration of mortar joints by manganese-colored brick. Am. Ceram. Soc. Bull. **45**, 271–76 (1966).
25. Brownell, W. E., J. L. Kenna, and P. P. Witko, Jr.: Staining of mortar by manganese colored brick. Am. Ceram. Soc. Bull. **45**, 1055–59 (1966).
26. Brownell, W. E.: Preparation of a nonbleeding manganese stain. Am. Ceram. Soc. Bull. **54**, 193–94 (1975).
27. Soderstrum, W. K.: Cleaning Clay Masonry, SCPRF (now Brick Inst. of America) McLean, Va., May 1964.
28. Young, J. E.: Some factors affecting the development and removal of vanadium efflorescence. Am. Ceram. Soc. Bull. **38**, 260–63 (1959).
29. Anonym: Efflorescence, Part 2: The Cleaning and Maintenance of Brickwork, Clay Prod. Bull., New Zealand Pottery and Ceramics Res. Assn., Pub. No. 19, August 1961.

12. Future Trends

12.1. Production

In any attempt to predict future events, it always seems proper to look at the immediate past for the purpose of extrapolating the trends from the past into the future. There are pitfalls in this procedure, as there are in all bases of prediction, but a review of the past is probably as good a place as any to start. At least if past trends are to change, we know that some basic patterns must change; and if such changes are a part of the predictive process, they must be justified. If the driving forces of the past remain unchallenged, then the trends of the past can be a satisfactory guide to the future.

The past trends in the use of structural clay products in the construction industry in the United States are presented in a particular way on Fig. 120 [1]. The percent of dollar value of all structural clay products used in new building was calculated from data provided by the U.S. Bureau of the Census and plotted yearly from the conclusion of World War II to 1971. During this period the economy of the United States was expanding, and free from such catastrophic events as seriously disrupt the structural clay products industry, as described in Chapt. 1 and illustrated in Fig. 7.

Face bricks were selected to examine in a way similar to that just presented for the industry as a whole in both residential and nonresidential building construction, because brickmaking is the largest segment of the structural clay products industry, and has been expanding in total dollar value and units produced over the same period of time. The relations between the value of face bricks produced and the values of all residential and nonresidential construction are also shown on Fig. 120. It can be seen that the value ratio of face bricks to that of building construction roughly parallels that of the whole structural clay products industry.

Fig. 120 shows that the dollar value of all structural clay products, and face bricks in particular, have not kept pace with the total value of building construction since 1947. The proportional value of structural clay products going into new building in the United States has decreased from around 2% in the late 1940's to around 1% in 1970.

These trends can be caused by two different behavioral patterns, or a combination of both. Either structural clay products have not moved up in price commensurate with other building materials—in which case they are now a real bargain—or they have been slowly giving way to other materials such as lumber, metals, plastics, and concrete products. If bricks have been held at a price that is too low, perhaps panelling or some other control over the productivity of labor will provide the extra profit margin that would be due the industry. If the competition of other materials is the primary factor, the economics of the future and a concerted research and developmental effort to increase real productivity, as defined by Kendrick [2], may be helpful factors to the whole structural clay products industry.

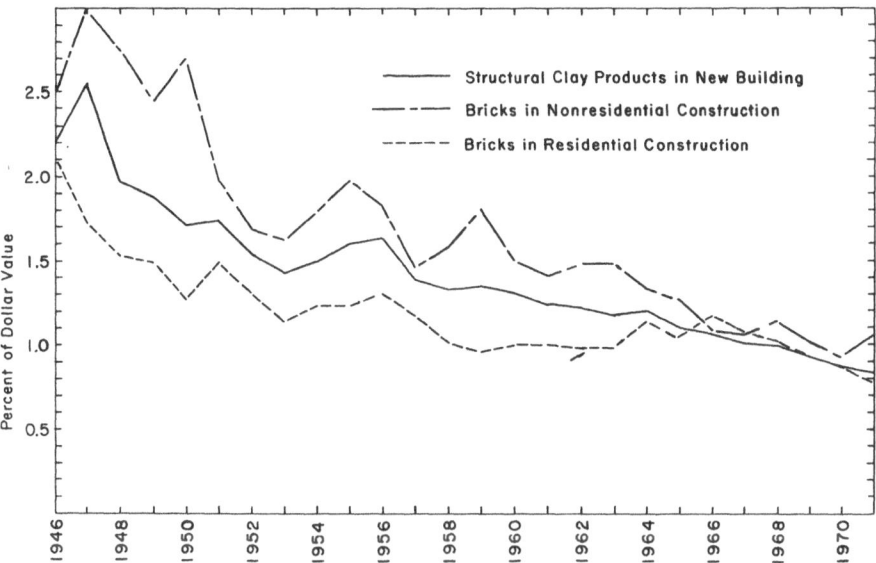

Fig. 120. Percent of dollar value of structural clay products going into new building since World War II (farm buildings not included) [1]

The actual volume of the various structural clay products produced yearly from 1946 to 1973 are given in Table 16. The production of face bricks has expanded over this period from around 5 billion to about 9 billion annually. Floor and wall tiles have increased from around 100 million-square feet annually to about 300 million. Sewer-pipe production doubled during the ten-year period from 1946 to 1956; however, it has fallen off slightly over the past 20 years. Glazed and unglazed facing tiles held a production level around 450-million standard brick equivalents from 1950 to 1959, but since then there has been a catastrophic decline in production to about 100 million equivalents. This decline in the use of structural facing tiles was probably due to the high unit costs and the encroachment of the more progressive face-brick industry, which now produces larger shapes that they call "Jumbo," "Through-the-Wall," and "SCR" bricks. Fired-clay partition tiles (unglazed, nonfacing units) were already under severe competition from concrete blocks by 1946, and the data in Table 16 show that this industry has been drying up since then—probably never to rise again as a unique and separate branch of the structural clay products industry. Other structural clay products that were essentially lost from the industrial scene in the United States before 1946 were terra cotta, roofing tiles, and paving bricks. These

Table 16. Production Trends in Structural Clay Products[1]

Date	Bricks (billions)	Facing Tiles[2] (million brick equiv.)	Partition Tiles (million short tons)	Sewer Pipes (million short tons)	Floor-Wall Tile[2] (million sq. ft.)
1946	4.869	274	1.273	1.081	61.5
1947	5.142	365	1.237	1.395	111
1948	5.842	335	1.263	1.496	104
1949	5.524	370	1.358	1.463	99
1950	6.333	434	1.294	1.549	125
1951	6.625	485	1.239	1.688	147
1952	5.889	413	0.976	1.649	132
1953	5.874	456	0.990	1.655	137
1954	6.720	432	0.981	1.763	178
1955	7.902	481	0.935	2.112	233
1956	8.085	497	0.862	2.154	246
1957	6.658	422	0.687	1.836	212
1958	6.489	446	0.574	1.773	222
1959	7.336	414	0.551	2.025	259
1960	6.943	392	0.496	1.955	242
1961	6.682	417	0.479	1.831	229
1962	6.889	393	0.439	1.680	258
1963	7.414	359	0.357	1.835	274
1964	7.878	337	0.346	1.908	295
1965	8.212	321	0.291	1.819	287
1966	8.384	324	0.246	1.705	284
1967	7.095[3]	233[3]	0.234	1.572[3]	258[3]
1968	7.557[3]	214[3]	0.193[3]	1.706[3]	275[3]
1969	7.290[3]	203[3]	0.242[3]	1.784[3]	285[3]
1970	6.496	169	0.181	1.622	250
1971	7.570	154	0.158	1.718	276
1972	8.402[3]	128[3]	0.101[3]	1.718[3]	308[3]
1973	8.923[3]	123[3]	0.094[3]	1.638[3]	301[3]

[1] U.S. Bureau of the Census, Statistical Abstract of the United States: 1948–74, Washington, D.C.
[2] Includes glazed and unglazed ware.
[3] Data are for shipments instead of production.

production trends in the various products of the industry encourage predictions that face bricks, floor and wall tiles, and sewer pipes will flourish here in the years to come while facing tiles and partition tiles will become extinct.

Whatever is happening to the proportional value of structural clay products going into the building industry, it needs correcting. There are no better building materials for long life and low maintenance. Bricks and tiles provide a high comfort factor for buildings because of their high heat capacities which ultimately means lower energy costs to heat and cool buildings. Such structures are fireproof, with no sacrifice of esthetics and with extensive variability in colors and textures. This country can no longer afford to overlook the very appropriate qualities of structural clay products for construction.

We have been living in what might be called a wasteful economy in the United States since 1946. After being denied the available standards of living through the 1930's, and willingly sacrificing them during World War II, the majority overreacted to the apparent abundance of the good things of life from 1946 to 1973. Food was wasted at the dinner tables, automobiles were built to last only a few years, household appliances were covered by only short warrantees, much clothing was discarded because of changing fashions, and energy sources were cheap and seemingly inexhaustible. So many buildings, both residential and nonresidential, were constructed to last no more than 50 years. In these cases the long-lasting qualities of structural clay products may have been ignored for the whims of variation regardless of cost.

In spite of the less-than-glamorous economic trends of the recent past in the structural clay products industry, it is inconceivable that a 10,000-year-old industry that has progressed into the climate of modern technology could casually disappear from the industrial scene.

There are several reasons for justifying a prediction that structural clay products will hold their place in the construction industry, and even gain in prominence in the future. Raw materials for structural clay products are inexhaustible and cheap, and these conditions are not likely to change. (It is quite possible that suitable clay deposits are forming on the earth faster than they are being used by this industry.) The processing of structural clay products is a relatively inexpensive operation, and there is ample room for improvement of productivity. The industry has proven over the centuries that it has the resiliency to rebound from economic disasters. Now it must withstand a more insidious erosion of its position in the construction industry.

In the near-future, certain competitive materials for building may become more expensive and less desirable. Metals and petroleum-based products are apt to become increasingly expensive because of short supplies. Concrete products will remain a stiff economic competitor, but fired clay products could be made cheaper and, at the same time, retain their desirability from the standpoints of dimensional stability, strength, attractiveness, and long-range durability. Lumber will continue to be a competitive material in individual home construction; however, costs may force more and more people into multiple-dwelling units where the need for fireproofing, durability, and insulation will favor structural clay products.

Face bricks can be more competitive if the productivity of building walls can be increased. Factory-fabricated panelling is the current approach to improved productivity, but large interlocking units employing a different bonding system may very well be more acceptable in the future. The conventional mortar joints have always been in question, as to whether the mortar acts to keep the bricks apart or to bond them together! A new approach to ceramic panelling has already been developed in the laboratories in which lightweight, decorated panels would be free from the appearance of brick walls [3, 4]. This system will be exploited in the future, and it will create a new industry to replace the already-lost terra cotta industry.

The floor- and wall-tile industry in the United States seems to be expanding and healthy. With continued research and development to increase productivity, it should continue to enjoy prosperity for the foreseeable future. This industry has done much and will do more to lower the overall installation costs.

Sewer pipes are the best product for the purpose intended. The present concerns for waste disposal without polluting the environment will demand larger and more sophisticated sewer systems throughout the land. The sewer-pipe industry should make noticeable gains in the next 30 years or so, because the protection of the environment is a real problem, not a passing fad. The availability of clean waters is vital for our survival as a progressive nation.

The years 1974 and 1975 have brought shocking realizations that we have been living in a wasteful economy and that it must change. Thoughts are being directed towards conserving natural gas, petroleum, metals, and food. It is likely that a more frugal economy will develop in this country over the next few years, and if this happens, there will be a return to constructing buildings that can be expected to last and be useful considerably longer than at present. When this concept becomes ingrained into our economy, there will be more emphasis on the use of structural clay products in building construction.

12.2. Technical Changes

The cost and limited availability of natural gas and fuel oil for drying and firing structural clay products is a current concern in the industry, and these manifestations of a nationwide energy problem are the first signs of the most pressing technical change in the future. The industry has been plagued by the short supply of natural gas since about 1950; so in many locations dual-fuel burners have been a necessity to maintain year-around operation. Now the supply of natural gas is more critical. No expansion in its use is permitted, and many factories have been cut off from this source of fuel. New installations and old have turned to fuel oil as a source of energy, only to find that the extra demands have created a shortage of this fuel as well. The short supply of natural gas and fuel oil has forced the prices of these commodities up recently, and the costs are certain to escalate in the near future.

Engineering research has started in a few places, in the last year or so, to investigate the use of powdered coal in firing ceramic kilns. These experiments have shown that the use of powdered coal has promise for the high-temperature zones of modern tunnel kilns, but gas or oil still seems to be essential for the low-temperature zones. Coal is in adequate supply, but the nationwide increase in demand for coal has suddenly sent the prices of this fuel soaring also. As a matter of fact, the early results of the experimentation in the use of coal to fire structural clay products have shown that the cost of this fuel is much higher than either gas or oil at this time. It appears that a return to coal as a primary fuel for modern kilns is only a stop-gap measure to cope with the energy crisis until something better has been worked out nationally. In the short-range future, perhaps through the next 20 years, fuel oil and powdered coal will be the primary fuels for existing installations. No doubt some use of gas manufactured from coal will also be used by some elements of the industry, but the cost is apt to prohibit general use.

In the long-range future complete conversion to electric energy for drying and firing structural clay products is inevitable. By 1990 electricity will be the most available and dependable source of energy for the industry. Because the costs of gas and oil will rise more rapidly than electricity, and because of the technical problems encountered with the use of coal, electrical energy will be the cheapest fuel by the year 2000. The abandonment of hydrocarbon fuels by the industry will present problems in maintaining color lines and high quality products. These problems will involve atmosphere control, but most of the basic science necessary to solve them is already known. Applied and engineering research will be required to produce economically the products desired in electrically fired kilns. This work should be started now, because a 20-year lead time is not excessive for changes of this magnitude.

The most promising kiln design for the introduction of electric firing is the continuous roller-hearth. With this type of kiln, fast firing will be possible, since the ware will not be stacked in large settings, and temperature variations across the ware will be minimal. Any continuous kiln can be designed to be an excellent heat exchanger, which will conserve energy by efficient utilization of the heat produced. These future kilns will be better insulated than those in current operation, to further conserve energy. In addition, low-setting, roller-hearth kilns greatly simplify automation, which is a rather complex problem for the present factories. The capital investment in kilns of the roller-hearth type, with the necessary atmosphere controls and heat exchangers, will be justified by the more efficient use of a continuously available energy source.

Some other technical changes that may come about in the future have to do with increase in productivity. Automation has come to the structural clay products industry only recently, and even though the massive machines are impressive to watch, in many ways they still seem crude and cumbersome. It is very likely that when different kilns and dryers are adopted, plant automation will become simpler, more versatile, and more dependable. Forming operations for face bricks may include dry pressing [5] or hot pressing [6]; however, some engineering details remain to be worked out. All of the estimates that have been made on the capital investment and operating costs for a hot-pressing plant are lower than those currently encountered. Another feature of the pressing operation is the greater flexibility in the raw materials that can be used. The restrictions imposed by the clay mineral content on plastic forming operations are removed. It is likely that the use of high-frequency electric drying will increase productivity in the future, when efficient generators have been developed for the evaporation of water from plastically formed bodies.

The third area in which technical changes will occur in the structural clay products industry is the improvement of quality with regard to durability in adverse environments. Further studies into pore-size distributions and microstructure may very well lead to the development of more universally durable products. Moisture expansion is becoming better understood through the efforts of various research programs, and proper compositions for glazed products should be forthcoming. The future may find the structural clay products industry more selective of raw materials, to provide better flexibility in operation, as well as more uniform products [7].

12.3. Research of the Future

Research to improve quality, increase productivity, and lower construction costs of the structural clay products has been fragmented and sporadic in the United States. From the 1930's into the 1950's our Bureau of Standards conducted research programs in support of the industry. For a very short period of time in the 1950's and 1960's, a cooperative industrial research effort was made by the face-brick industry. The most continuous research effort has been made by the sewer-pipe industries in a cooperative venture that has lasted since 1949. The research efforts of the face-brick, structural-tile, and floor- and wall-tile industries are now carried out in very limited ways by five or six of the larger corporations. In contrast, many other countries have continuous, government-sponsored building-research programs that have contributed a great deal of information for worldwide use.

The reasons for the lack of effective research and development in the structural clay products industry in our country are three-fold. First, the industry grew from antiquity by means of many small enterprises that have little wealth to invest in research. Second, the conservatism in the industry resists change and innovations which, in turn, probably stems from the conservatism inherent in the building industry. Third, our government research facilities are no longer attempting to assist private industry by way of research, even in cases where the industries need a centralized effort.

Basic and applied research programs are essential to progress, growth, and increased productivity. This has been proven over and over again by the newer and more progressive industries. Somehow or other the structural clay products industry must engage in a never-ending research effort where a proportion of total sales is always consigned to basic research. In this country, this will be done through the large corporations which are expanding in both size and numbers, and as mentioned in Chapt. 1, there has already been a case where large corporations have united their resources on a common project. What is really needed in this industry is *direction* which will point toward the essential problems, avoid duplication, and provide a continuous effort, almost regardless of the state of the economy. The kinds of research and development considered here are broader and more basic than the solution of specific intraplant problems. (These will always have to be handled by the plant engineers and scientists.)

Some of the areas into which future research must go, in order to place the structural clay products on a firm foundation, are listed below. It is hoped that these topics will stimulate interest and action by companies, institutes, and universities. Good work in these areas will produce increasing productivity which will, in turn, make the industry more competitive, provided that the companies are willing and have confidence to make the necessary changes.

Fundamental Research
Stress-strain relations in plastic clay bodies
Drying behavior of clay masses
Pore structure of fired clay bodies
Microstructure of fired clay bodies
SO_3 adsorption on silicate surfaces
Compatibility of fired phases to reduce microcracks

Engineering Research
Microwave drying of clay wares
Improved face brick design to simplify wall construction
Development of better bonding materials
Production of low-cost partition and foundation tiles or bricks
Electric firing in production

12.4. Summary

The structural clay products industry produces high-quality products for the construction industry. For the most part, face bricks and tiles have adequate durability, and they are presented to the customers in a multitude of sizes, colors and textures [8]. The sewer pipes manufactured in the United States and Canada are excellent products and are constantly improving. The prices charged for all of these products at the factories appear to be a good value for the construction industry. In spite of these fine qualities, the structural clay products industry does not seem to be maintaining its share of the total building market.

Productivity, i.e., units produced per man-hour or per dollar of capital invested, must increase in the United States to make the industry more competitive with other, less desirable, products. It appears that this can be done by developing better raw materials, by appreciating their behaviors, and by tieing all processes together with simpler and faster automation. The plants of the future will operate all departments 24 hours a day, 7 days per week [9].

The structural clay products industry must quickly solve its energy-consumption problem. While short-term solutions with hydrocarbon fuels is imperative, the industry should expect to convert to electrical power for all phases of manufacturing. Conversion to electric drying and firing will require more attention to atmosphere control. New kiln designs with better heat exchange and more insulation will be necessary, and it seems that the multitiered, roller-hearth kilns should be carefully considered.

High product quality needs greater uniformity across the industry. Occasionally a faulty product will reach the market, quite unintentionally, due to lack of knowledge by those in charge of production. When this happens, the displeasure of the customer casts a long shadow over the whole industry. Difficulties of this kind can be overcome by continued research related to durability against moisture, exterior salt contamination, and freezing temperatures.

The structural clay products industry in the United States must shake off its conservatism and progress toward stabilizing and improving its position in the construction industry.

References

1. U.S. Bureau of the Census, Statistical Abstract of the United States: 1948−74, Washington, D.C.
2. Kendrick, J. W.: Postwar Productivity Trends in the United States, 1948−1969, Columbia Univ. Press, New York, 1973.
3. Coffin, L. B., and W. G. Lawrence: Lightweight Building Materials. Mon. Prog. Rept. No. 311, New York State College of Ceramics (May 1962).
4. McMahon, J. F.: New Family of Lead-Glazed Lightweight Structural Products. Mon. Rept. No. 348, New York State College of Ceramics (June 1965).
5. Schmidt, E. W.: Dry-pressing in the brick industry. Sprechsaal **102** (2−3), 62−77 (1969).
6. Brownell, W. E., F. C. McMann, and S. D. Jang: Hot pressing of face brick: III Pressing parameters for quality products. Am. Ceram. Soc. Bull. **53**, 440−42 (1974).
7. Earl, W. A., and W. E. Brownell: Composition of an ideal face brick. Am. Ceram. Soc. Bull. **42**, 49−51 (1963).
8. Svec, J. J.: Construction and clay products: foundation for prosperity. Brick Clay Rec. **163** (4), 22 (1973).
9. Svec, J. J.: Leitl-Werke: First continuously operating plant. Brick Clay Rec. **163** (6), 22−24 (1973).

Subject Index

Minerals, disilicate 24
—, essential *36*
—, nonessential *37*
Mining *43*
— pollution *50*
— procedures *47*
— with draglines 48
— — front-end loaders 48
— — scrapers 48
— — shovels 48
Mississippian period 44
Modulus of elasticity 161
— — rupture 111, 112, 161
Moisture control 88, 89, 191
— differential 110
— expansion 15, 16, 142, *203*, 218
— gradient 101–104, 107, 114, 119
— stress 104, 105, 111, 112
Molds 3–5, 20, 83–85, 98
Momence, Ill. 8
Monhenjo-daro 1
Montezuma, Ind. 8
Montmorillonite 13, 31, 34, 35, 37, 38, 40, 70, 109, 110, 112, 131
Mortar 175, 205–210, 216
Mount Savage, Md. 8
Mullite 128–131, 135, 141, 142, 150, 203
— nucleation 134
Muriatic acid (see hydrochloric)
Muscovite 31, 34, 35, 38, 40, 45

Nacrite 31
Napthalene diisocyanate 184
National Clay Pipe Institute 16, 176
— Fireproofing Company 11
— Terra Cotta Society 16
Neolithic 2
Nephelene syenite 172
New England Brick Company 11
New Jersey 8, 11, 14
New York City 6
New York State 44
New York State College of Ceramics 14, 17
Nile Delta 4
Nineveh 3
Nippur 3
Nitrile rubber 180
Nodules 39
Nonplastics 101
— fraction 56
Nontronite 31, 34
Normal distribution 60
North Carolina 6
Northwestern Terra Cotta Co. 16

Oaks Factory 10
Octahedral sheets 30, 34–36, 130
— site 28
Octahedron 26
Ohio 11
Oligoclase 38, 40
Opacifiers 173
Ordovician 44
Organic matter 44
Orientation, particle 79, 81
— of clay grains 80, 92
—, preferred 93, 110, 112
Orthoclase 38
Orthosilicates 38
Overfiring 135
Oxalic acid 210
Oxidation 155
— of carbon and pyrite *145*
— -reduction *144*
Oxygen 143, 145, 146
—, electronic structure of 64

Packaging 17, *176*, 197
Paint stains on brickwork 210
Pakistan 1
Paleozoic era 44
Pallets 84, 85, 97
Panelling 213, 216
Panels, factory fabricated 17, 20, *175*, 179, 205, 215
Particle size 101, 109
— -size analysis 57, 69
— -size distribution 53, *55*, 103, 110, 111, 166, 188, 190, 191, 206
Penetrameter test 191
Pennite 31
Pennsylvania 6, 11
Pennsylvanian period 44
Periclase 28
Permian period 44
Persulfates, metal 180
Perth Amboy, N.J. 8
— Amboy Terra Cotta Company 8
Peru 1
Petrographic microscope 13
Pharaoh Sheshouk I 4
Phase diagrams 133
Phi scale 58, 59, 200
Phlogopite 31, 35
Phoenixville, Pa. 10
Pigment 165, 209
Pipe, chemical resistant 19, 203
—, clay 4, 19, 86, 90, 178, 179, 198
—, conduit 19
—, crushing strength of 203